S0-AWV-392

Pesticide Waste Management

ACS SYMPOSIUM SERIES **510**

Pesticide Waste Management

Technology and Regulation

John B. Bourke, EDITOR
Cornell University

Allan S. Felsot, EDITOR
Illinois Natural History Survey

Thomas J. Gilding, EDITOR
National Agricultural Chemical Association

Janice King Jensen, EDITOR
U.S. Environmental Protection Agency

James N. Seiber, EDITOR
University of California—Davis

Developed from a symposium sponsored
by the Division of Agrochemicals
at the Fourth Chemical Congress of North America
(202nd National Meeting of the American Chemical Society),
New York, New York,
August 25–30, 1991

American Chemical Society, Washington, DC 1992

Library of Congress Cataloging-in-Publication Data

Pesticide waste management: technology and regulation / John B. Bourke, editor . . . [et al.].

p. cm.—(ACS symposium series, ISSN 0097–6156; 510)

"Developed from a symposium sponsored by the Division of Agrochemicals at the Fourth Chemical Congress of North America (202nd National Meeting of the American Chemical Society) New York, New York, August 25–30, 1991."

Includes bibliographical references and indexes.

ISBN 0–8412–2480–3

1. Pesticides—Environmental aspects. 2. Pesticides—Containers. 3. Pesticides—Law and legislation—United States.

I. Bourke, John B., 1934– . II. American Chemical Society. Division of Agrochemicals. III. American Chemical Society. Meeting (202nd: 1991: New York, N.Y.) IV. Chemical Congress of North America (4th: 1991: New York, N.Y.) V. Series.

TD196.P38P48 1992
363.17′925′0973—dc20

92–32304
CIP

The paper used in this publication meets the minimum requirements of American National Standard for Information Sciences—Permanence of Paper for Printed Library Materials, ANSI Z39.48–1984. ♾

PRINTED IN THE UNITED STATES OF AMERICA

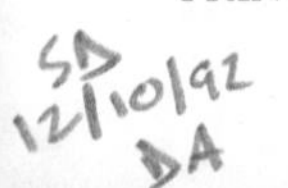

1992 Advisory Board

ACS Symposium Series

M. Joan Comstock, *Series Editor*

Foreword

THE ACS SYMPOSIUM SERIES was first published in 1974 to provide a mechanism for publishing symposia quickly in book form. The purpose of this series is to publish comprehensive books developed from symposia, which are usually "snapshots in time" of the current research being done on a topic, plus some review material on the topic. For this reason, it is necessary that the papers be published as quickly as possible.

Before a symposium-based book is put under contract, the proposed table of contents is reviewed for appropriateness to the topic and for comprehensiveness of the collection. Some papers are excluded at this point, and others are added to round out the scope of the volume. In addition, a draft of each paper is peer-reviewed prior to final acceptance or rejection. This anonymous review process is supervised by the organizer(s) of the symposium, who become the editor(s) of the book. The authors then revise their papers according the the recommendations of both the reviewers and the editors, prepare camera-ready copy, and submit the final papers to the editors, who check that all necessary revisions have been made.

As a rule, only original research papers and original review papers are included in the volumes. Verbatim reproductions of previously published papers are not accepted.

M. Joan Comstock
Series Editor

Contents

RINSATE MINIMIZATION AND REUSE

CURRENT DISPOSAL TECHNOLOGIES

Preface

THE OCCURRENCE OF PESTICIDE RESIDUES in surface water and groundwater has been documented in a plethora of research published in a wide variety of scientific journals and books over the past 30 years. Pesticide residues in surface water and groundwater originate as point or nonpoint sources. Point sources are defined as those arising from routine handling procedures, spills, and waste disposal associated with commercial operations, which generally include agrochemical retail dealerships, manufacturing facilities, warehouses, and means of transportation. Nonpoint sources of pesticide contamination are diffuse, arising from the practices of many farmers during the course of routine operating practices, which traditionally include application of pesticides to the field and tank loading, mixing, and rinsing. Ironically, tank loading, mixing, and rinsing on the farm lead to the same problems as handling operations at commercial facilities—pesticide residues at high concentrations in localized areas with the potential of spreading to water resources. Regardless of the semantics used in defining the origin of pesticide contamination, all practices related to generation of spills and waste, whether they are farm or industry related, can be managed. Failure to implement waste management has resulted in high levels of pesticide residues in soil and water at pesticide-handling sites. This problem is by no means new. Two previous ACS Symposium Series books and several U.S. Environmental Protection Agency publications have addressed many pesticide waste-disposal issues.

How much have we progressed since publication of the last ACS Symposium Series book on pesticide waste in 1984? The report card is mixed; although much progress has been made in recycling containers and minimizing wastewater, we are only beginning to come to terms with past uncontrolled disposal practices at agrochemical retail facilities that have caused high levels of soil and groundwater contamination. However, we can be optimistic about future cost-effective technologies for small businesses and farms because research on innovative treatment of wastewater and unused pesticides is progressing.

In keeping with the long-standing interest of the ACS Division of Agrochemicals in the safe use of pesticides and their effect on society and the ecosystem, this book presents the current status of pesticide waste-management technology. It goes one step further than previous publications because it discusses pesticide waste regulations and implementation of these regulations from the viewpoint of several regulatory agencies. In addition to presenting eight chapters on current disposal technologies for wastewater and soil clean-up, the book includes seven chapters on container recycling and disposal, two chapters on alternative application technologies for rinsate minimization, and three chapters on remediation of contaminated sites.

The symposium on which this book is based was meant not only to inform researchers of progress in pesticide waste-disposal issues but also to produce useful information for state and federal officials who have to grapple daily with problems created by pesticide waste and with regulatory enforcement. After reading the chapter on problems of waste management in developing countries, we can conclude that the United States is doing a good job. We sincerely hope this book is useful beyond the research phase and will serve as a stimulus for action among all parties using or regulating pesticides.

JOHN B. BOURKE
Analytical Chemistry Laboratory
Cornell University
Geneva, NY 14456–0462

ALLAN S. FELSOT
Illinois Natural History Survey
Champaign, IL 61820

THOMAS J. GILDING
National Agricultural Chemical Association
1155 Fifteenth Street, N.W.
Washington, DC 20005

JANICE KING JENSEN
Office of Pesticide Programs
U.S. Environmental Protection Agency
Washington, DC 20460

JAMES N. SEIBER
Center for Environmental Sciences and Engineering
University of Nevada—Reno
Reno, NV 89557

February 3, 1992

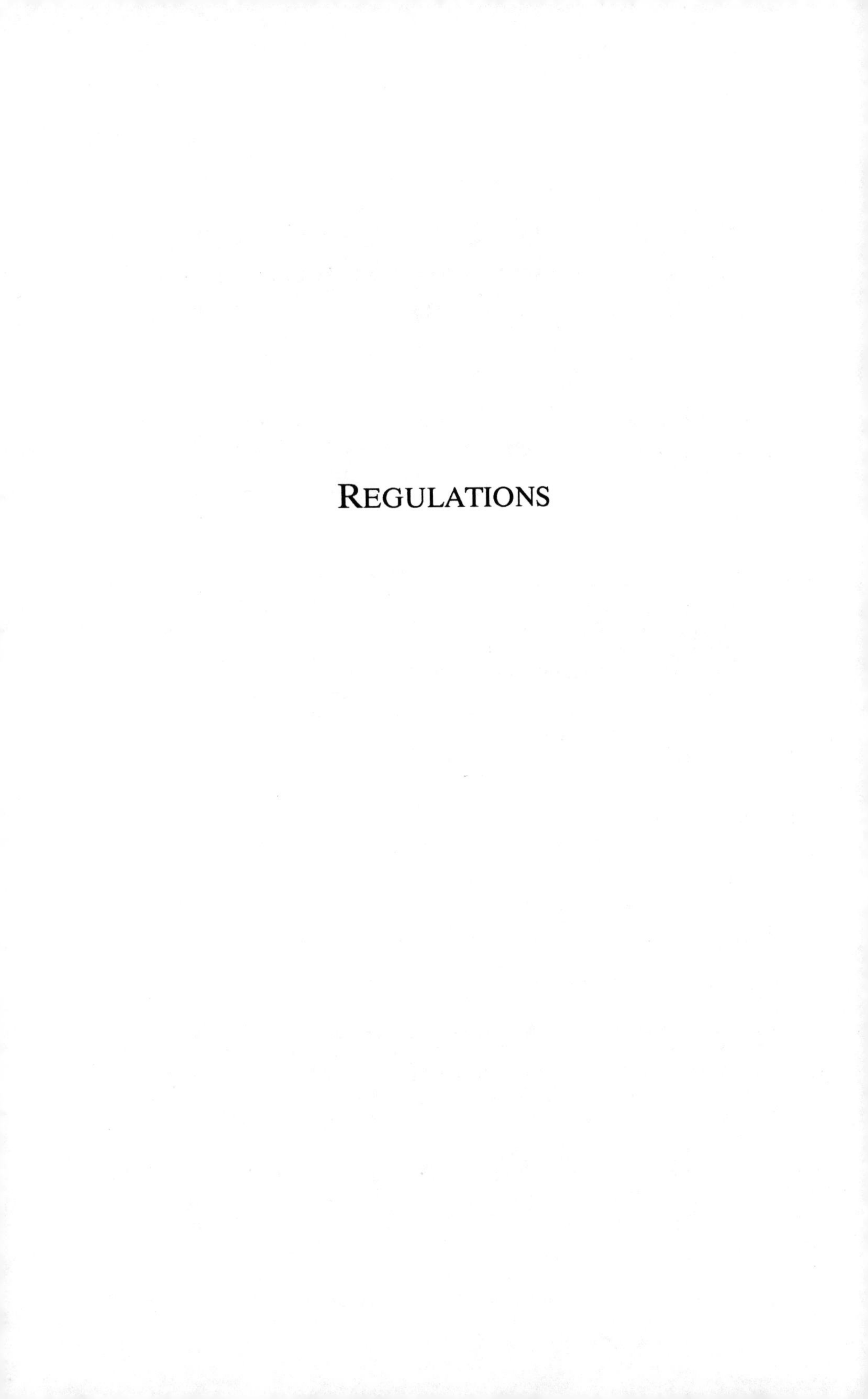

REGULATIONS

Chapter 1

Pesticide Container Regulations as Part of the U.S. Environmental Protection Agency's Strategy

Nancy Fitz

Office of Pesticide Programs (H–7507C), U.S. Environmental Protection Agency, 401 M Street S.W., Washington, DC 20460

The United States Environmental Protection Agency (EPA) is currently revising the pesticide container regulatory scheme. In 1988 Congress reauthorized the Federal Insecticide, Fungicide, and Rodenticide Act, which requires EPA to address pesticide containers in three ways: (1) to conduct a study of pesticide containers and report the results to Congress; (2) to promulgate container design regulations; and (3) to promulgate residue removal regulations. The recommendations for container rinsate and empty container management from the pesticide disposal workshops held during the 1980s are reviewed. In this context, the EPA container management strategy and the basic philosophy of the draft regulations are presented. Several unresolved pesticide container issues are then discussed.

Since the 1988 amendments to the Federal Insecticide, Fungicide, and Rodenticide Act (FIFRA), the U.S. Environmental Protection Agency (EPA) has been investigating pesticide container issues. This paper describes the current EPA container management strategy, including the basic philosophy of the regulations being drafted. The EPA approach is put into context by reviewing the recommendations for containers from the mid-1980 disposal workshops. Additionally, some problems that will continue to be issues in the future are discussed.

The Past: Where We Were

Managing Pesticide Wastes: Recommendations for Action (*1*), a summary of the National Conferences and Workshops on Pesticide Waste Disposal, suggests pesticide waste management goals for all groups involved with pesticides. The document presents recommendations for action in several categories, including management practices and regulations. The management practice recommendations for dealing with empty containers, in order of preference, are:

- Container minimization;
- Container reconditioning/recycling; and
- Container disposal, which includes (1) proper rinsing, (2) collection programs, and (3) sufficient disposal options.

These recommendations are generally aimed at registrants, states, and pesticide users, with some overlap with the regulatory recommendations.

In its regulatory recommendations, Managing Pesticide Wastes defines empty pesticide containers as one of five categories of wastes that should be regulated by FIFRA. The document recommends a core for the regulatory structure that is based on a series of Waste Management Practices (WMPs), which are generally acceptable procedures for storing and disposing of wastes.

Container Rinsate. The recommendations for container rinsate include the following.

- "Waste Management Practices (WMPs) should establish triple rinsing as the minimum mandatory requirement for rinsing empty pesticide containers and should identify the exact procedure for triple rinsing.
- "Pressure rinsing should be encouraged over triple rinsing. To ensure acceptable performance for pressure rinsing, minimum ranges for water pressure and rinsing time should be specified.
- "Applicators should be required to rinse containers at the time they are emptied and to drain the rinsate into the pesticide mix tank.
- "If water is not acceptable as the diluent for rinsing the container of a specific pesticide formulation, then the registrant should be required to identify the correct diluent on the product label." (*1*)

Empty Containers. In addition to the above recommendations for rinsing the containers, the document makes the following recommendations for empty containers.

- "Non-mandatory WMPs for container rinsing could address such needs as procedures beyond the specified mandatory triple rinsing procedure and optional rinsing-equipment designs.
- "WMPs for container collection programs should be limited to general mandatory prohibitions concerning the location of collection sites (to protect human health and the environment) and storage security of containers. Container rinsing should not need to be addressed for off-site container collection sites, since rinsing of containers would be mandatory at the time of emptying at the mix site.
- "Container reuse and recycling - for metal scrap or energy value - should be encouraged, but not mandated, over discarding or destruction.
- "The WMPs could provide significant technical assistance and design standardization in reusable container concepts.
- "Land disposal of empty pesticide containers must be maintained as a 'backup' disposal option for the near-term, but should be phased out as recycling and incineration become acceptable and accessible. The WMPs should define general mandatory requirements for land disposal sites, primarily to preclude the use of locations where ground water and surface water are at risk. Normally, landfills approved for industrial and municipal solid wastes should be acceptable for disposing of empty pesticide containers.
- "WMPs should establish acceptable procedures for burning combustible containers at the mixing site. The procedures should identify the type and number of containers considered acceptable for on-site burning and should establish mandatory requirements for burning to ensure adequate protection of health and the environment. States should have the discretion of adopting the procedures for use in their own FIFRA waste management programs. Open burning should not be made applicable to homeowners." (*1*)

EPA is making significant steps in fulfilling many of these recommendations in the current regulatory effort, particularly those relating to rinsing. On the other hand, some of these issues remain unresolved.

The Present: Where We Are

In 1988 Congress amended FIFRA, which now requires EPA to address pesticide containers in three ways: (1) to conduct a study of pesticide containers and report the results to Congress; (2) to promulgate container design regulations; and (3) to promulgate residue removal regulations. These projects are very interrelated because the information collected during the study is used in the report (*2*) and as support for the regulations being drafted.

Container Study. During the past several years, EPA has gathered the available information on pesticide containers through a variety of methods. Four open meetings were held with representatives from different perspectives, including pesticide manufacturers, pesticide packagers, state agencies, container manufacturers, environmental groups, and trade and user associations. Follow-up meetings were held with many of the participants to discuss specific issues in greater detail. Additionally, EPA staff members have made several field trips to meet with growers, applicators, dealers, and distributors to increase the Agency's "real world" knowledge of container handling practices.

In conducting the study, EPA has distinguished two major types of pesticide containers -- nonrefillable and refillable -- with substantially different concerns and issues for each type. Nonrefillable containers are generally considered one-way or throw-away packages and include most drums, cans, jugs, bags, bag-in-a-box designs, and aerosol cans. Refillable containers are those containers specifically designed to be refilled and reused. Examples include bulk storage tanks, minibulks, refillable bags, and small volume returnable containers.

The report to Congress (*2*) summarizes and consolidates the existing knowledge and data on pesticide containers and current handling practices. The issues and current practices regarding use, residue removal, and disposal are discussed for both nonrefillable and refillable containers. Also, the EPA's approach to managing containers is described with options and suggestions for further study.

EPA Container Management Strategy. Several general conclusions relating to the development of a pesticide container management strategy emerged from EPA's study. Managing Pesticide Wastes recommends that "FIFRA WMPs should address all aspects of empty container management, from time of emptying to time of disposal." EPA is taking an even broader approach and looking at the entire life of the container, including container integrity and transferring the pesticide from the container. Part of EPA's strategy includes promulgating the container design and residue removal regulations, which will be discussed in the following section. Additionally, the pesticide container management strategy includes long-term goals, which can be divided into several main categories.

View Formulation and Container as a Unit. The first long-term goal is to have the pesticide industry consider the pesticide formulation and its container as a single entity. The change in perception from considering a container simply as a vessel to transport a pesticide to seeing the container as an important part of the pesticide itself is an integral step in the long-term improvement of containers. Many phenomena, such as dripping, "glugging," and the residue in a container after it is cleaned depend on both the container and the formulation, as well as other variables. Therefore, the relationship between the container and the pesticide is important in all stages of the pesticide/container life cycle, including container use (transportation, storage, transferring pesticide from the container, etc.), residue removal, and container disposal.

Provide Leadership. The second long-term container management goal is to provide leadership in the area of pesticide containers. The container study involved a great deal of cooperation between EPA, other federal agencies, state agencies, industry groups, environmental organizations, and many individuals involved with pesticide containers. EPA would like to continue this dialogue and cooperation in the future.

Move toward Environmentally Preferable Containers. Another part of EPA's leadership role is to monitor and affect the trends of pesticide containers. In conducting the study, EPA determined that there are several desirable classes of containers. EPA has identified a hierarchy of environmentally sound container classes, which is presented below. This hierarchy is based on information collected on container use, residue removal, and container disposal, as well as the concepts of pollution prevention and reducing solid waste. The Agency would like to encourage the development and use of the most desirable container classes.

Within the hierarchy, the container classes are listed from most desirable to least desirable. For the purposes of this paper, a container is considered recyclable if the technology exists to recycle the material from which the container is constructed.

- Refillable containers and water-soluble packaging;
- Nonrefillable, recyclable containers that are currently being recycled;
- Nonrefillable, recyclable containers that are not currently being recycled; and
- Nonrefillable, non-recyclable containers.

This hierarchy is very similar to the recommendations for empty container management practices in Managing Pesticide Wastes, i.e., (1) container minimization; (2) reconditioning/recycling; and (3) environmentally sound disposal methods.

Refillable containers and water-soluble packaging are the most desirable container class because they support the concepts of waste minimization and pollution prevention. Specifically, these types of containers reduce or eliminate the need for residue removal and reduce the number of containers requiring disposal.

EPA realizes refillable containers and water-soluble packaging are not possible in every situation and that nonrefillable containers will always exist. The next category -- nonrefillable, recyclable containers currently being recycled -- is attractive because it reduces the number of containers requiring disposal as waste.

The third category, nonrefillable, recyclable containers not currently being recycled, includes most nonrefillable steel and plastic containers. With the proper infrastructure and market, the containers in this category could move up the hierarchy to reduce the number of containers requiring disposal.

The least desirable category, in terms of resource conservation, includes nonrefillable, non-recyclable containers. For example, because multiwall paper shipping sacks are usually constructed of more than one material (e.g., kraft paper and a barrier layer), they are not recyclable.

Container Design and Residue Removal Regulations. The EPA is currently drafting container design and residue removal regulations. In general, EPA is leaning towards performance standards, although some design standards are also being considered. Because the regulations are still in the draft stage and are subject to change, the potential requirements can be discussed only in general terms: safe use, residue removal, and disposal for both nonrefillable and refillable containers.

Nonrefillable containers. One problem that EPA is addressing for the use of nonrefillable containers is potential worker exposure while transferring pesticide from the container, i.e., if the container drips or "glugs." Options that the regulations may address include standardizing container closures to encourage the use of mechanical "closed" transfer systems and establishing performance standards to minimize dripping and glugging.

EPA's approach to residue removal from nonrefillable containers is proceeding along two tracks, one set of requirements for registrants and one for end users. Currently, most of the burden for residue removal is on the end user. One intent of these regulations is to increase the role of the registrants in residue removal considerations. EPA's goal is to improve the design of containers, based on the interaction between the container and the formulation, to facilitate residue removal.

One regulatory approach under consideration is to set a performance standard for the maximum amount of residue that remains in a container after a specified residue removal procedure is performed. A different standard could be set for each class of container, e.g., rigid containers with dilutable products or nonrigid containers. The registrant would be responsible for showing that the containers could meet this standard. The registrant could vary the container (size, shape, etc.) or the formulation in order to meet the standard.

On the other hand, it is the end user's responsibility to follow the label directions and properly clean the containers. Therefore, EPA is considering addressing residue removal at the end user level by defining standard procedures for both triple and pressure rinsing and by directing the user to clean the container immediately upon transferring the pesticide from the container.

One of EPA's major accomplishments has been developing data on residue removal. Standard laboratory protocols for triple rinsing, pressure rinsing, and emptying bags were developed with industry comment and a variety of container/formulation combinations were tested. The effects of variables in the triple rinse procedure, such as the initial drain time and shaking time, are now being studied in order to develop a quick, yet effective, triple rinse procedure. Research is planned to investigate variables involved with pressure rinsing, particularly the design of the pressure rinsing device.

In terms of the regulatory scheme for disposing of empty containers, Managing Pesticide Wastes states that "The primary purpose should be to ensure that containers are recycled or disposed of in an environmentally sound, yet practical, manner." EPA believes that the key to safely recycling or disposing of containers is to have clean containers. Having registrants ensure that the containers can come clean and making it easier for users to clean containers should facilitate the safe recycling or disposal of nonrefillable containers.

Refillable containers. While refillable containers offer several advantages in terms of waste minimization, they do present several new concerns. Specifically, the major problems EPA is addressing for refillable containers are the possibility of cross-contamination and the potential for larger releases of pesticide in the case of a container failure.

Some of the standards being considered to minimize the possibility for cross-contamination are:

- Using one-way valves and tamper-evident devices on liquid minibulk containers;
- Having the registrant develop residue removal procedures to be performed when a different pesticide is going to be put in the container; and
- Establishing a tracking system including serial numbers on the container and record-keeping.

Some of the options that are being considered to minimize the potential for container failures include:

- Developing integrity standards, such as drop tests, for each type of refillable container;
- Permanently marking the container with its date of manufacture; and
- Developing and performing a container inspection procedure.

Also, EPA is currently considering regulations for secondary containment (diking) around bulk tanks and for containment pads at areas where containers are regularly rinsed or refilled.

The Future: Where We Are Going

EPA has gained a great deal of knowledge about pesticide containers that has been extremely useful in pointing the way towards new regulations, policies, and goals. Significant progress also is being made toward fulfilling many of the regulatory recommendations in Managing Pesticide Wastes. However, many questions have been uncovered or remain unanswered and much work remains to be done on pesticide containers.

First, the container design and residue removal regulations that will go into effect over the next few years will require considerable effort from all of the groups involved. EPA will have to implement the regulations and develop training materials, while industry and end users will have to make some changes to comply with the new standards.

Second, problems with the disposal of nonrefillable containers will probably get worse before they get better. Very few disposal options are available to end users, and these are becoming more restricted. While this paper has not addressed disposal because it is not solely under the jurisdiction of FIFRA, several related points can be made.

Landfilling and open burning are the most common disposal methods for nonrefillable containers. An increasing number of landfill operators are refusing to accept triple-rinsed pesticide containers, even though they are not considered hazardous waste. The open burning of solid waste is banned by federal RCRA regulations. In addition, a number of state regulations specifically address the open burning of pesticide containers.

While significant progress is being made in developing the recycling of pesticide containers, we are still far away from having a national infrastructure for collection and an established market for the recycled material. Until this point is reached, the disposal of refillable containers will continue to be a serious problem.

Finally, while refillable containers and water soluble packaging offer several important advantages, they are not problem-free. EPA does not want simply to replace one problem (i.e., container disposal) with another one (i.e., larger spills or accidents). Therefore, these kinds of packaging need to be monitored closely as they continue to become more common.

An additional issue mentioned in Managing Pesticide Waste that needs to be seriously addressed is standardizing refillable containers. Some industry discussion has taken place, but little progress has been made on many of the issues.

EPA and industry have learned much about pesticide containers over the past several years. Many of the problems discussed at the 1980s disposal workshops are being addressed and progress is being made in solving them. However, there are many problems that have not been addressed, and some that are just being discovered. Pesticide container disposal will continue to be an issue in the 1990s.

Literature Cited

1. Managing Pesticide Wastes: Recommendations for Action, Summary of National Conferences and Workshops on Pesticide Waste Disposal, July 1988.
2. U.S. EPA, Office of Pesticide Programs, Pesticide Containers: A Report to Congress (Draft), July 1991.

RECEIVED April 21, 1992

Chapter 2

State Pesticide Disposal Regulations and Programs

Barbara B. Lounsbury

Legal Consultant, 505 West Auburn Road, Auburn, ME 04210

This paper briefly examines the status of state pesticide regulations and programs and finds that: (1) In most states many agencies and many statutes govern aspects of pesticide disposal, often resulting in real or perceived conflicts; (2) In many states the pesticide label constitutes the primary means of conveying information to the user and most enforcement activity depends upon label interpretation. Labels are not adequately drafted for these purposes; (3) Burning and burial of containers is allowed in some states although they are practices of questionable safety; (4) Emerging state regulatory and non-regulatory programs for pesticide container collection, waste pesticide collection and rinsate containment offer constructive ways of dealing with disposal issues; and (5) EPA and states could lessen the risk of environmental harm by changing label language, amending hazardous waste regulations, requiring transportation costs to be covered in any cancellation and suspension, coding containers as to date, dealing with cross contamination and management of sludges, expanding certification and training requirements, and ensuring that pesticide disposal issues are coordinated with waste reduction and recycling programs generally.

The current status of state pesticide disposal regulations and programs may be summarized in five broad statements: (1) Most states are caught in a web of multi-agency and multi-statutory jurisdiction; (2) Most states regulate non-agricultural pesticide wastes as part of the municipal solid waste disposal structure; (3) In many states, pesticide labels with minimal disposal instructions constitute the primary means of conveying information to the user and most state enforcement activity in these states depends upon label interpretation, (4) Emerging state regulatory and non-regulatory programs for container collection, waste pesticide collection and rinsate containment offer creative ways to address problems caused by pesticide disposal, and (5) Both states and EPA need to enact significant changes in regulations in order

0097–6156/92/0510–0008$06.00/0

to lessen the risk of environmental harm from pesticide disposal. This paper introduces these topics and suggests several issues that should be addressed and ways that practices could be improved.

Jurisdictional Complexity

Only six states house pesticide disposal under one environmental agency --Alaska, California, Connecticut, New Jersey, New York and Rhode Island. In the others, several agencies regulate the various issues and may not always agree on approaches to handle disposal issues.

Many state statutes affect disposal: air pollution laws, water quality laws for surface and ground water, solid and hazardous waste laws and waste reduction legislation , toxics substances legislation, hazardous substances transportation regulation, worker protection laws, Right-to-Know legislation, business practices legislation, fire codes, buildings codes and zoning regulations. Common law and liability concerns may also affect pesticide disposal practices.

Enforcement of the Label

The pesticide label is in many states the primary means of conveying information to the user and most state enforcement activity in these states depends on label interpretation. The Georgia Department of Agriculture recently wrote a letter to EPA on behalf of the State FIFRA Issues Evaluation Group ("SFIREG") emphasizing the critical role of labels:

...the label still remains the primary means of conveying information to the user and most of our enforcement activity depends on label interpretations. If labels are confusing, vague or misbranded, misuse may result; farmers may apply pesticides to the wrong crops, they may not use the right protective clothing for the desired use, they may improperly dispose of containers and rinsates, just to mention a few of the more serious consequences.(Georgia Department of Agriculture to Anne Lindsay, U.S. EPA, letter dated March 14, 1990)

There are two serious problems with this reliance on labeling. First, EPA and many states do not have readily accessible files of labels for all products currently registered or that were once registered and may still be in storage. Second and more important, state lead agency officials assert that , in many situations, labels do not provide them with satisfactory language for enforcement actions related to disposal and do not give applicators adequate guidance. They point to several flaws in labeling:

Labels do not contain sufficient information. In a recent survey by the Montana Department of Agriculture, 26% of applicators (300 responses) and 31% of dealers (130 responses) replied that pesticide product labels did not "provide adequate information to assist [them] with disposal." The Department of Agriculture concluded: "With this source of information significant for users, companies may need to look at clarifying information on the labels so all users consider the information adequate."(Montana Agricultural Business Association, Montana Aviation Trades Association, and Montana Department of Agriculture, <u>Summary of Pesticide Applicator and Dealer Disposal Survey</u>, November 1990.)

Labels may be ambiguous so that enforcement agencies decline to take action or applicators and agencies differ over what is and is not prohibited.

Labels may use terms that are confusing or misleading because of the use of these terms in other statutory schemes. For example, "...[T]he terminology between FIFRA and RCRA can be confusing to the pesticide user. A pesticide may be classed as highly toxic according to the FIFRA scheme...but not categorized as a toxic hazardous waste. Conversely, a pesticide may have a lower order of toxicity according to this scheme, but still be listed as a RCRA toxic waste...Language on the pesticide label can also be misleading. Each label contains directions for storage and disposal. Included on several pesticide labels is a statement that says 'pesticide wastes are acutely hazardous'; however, in most instances the chemical of concern is not by definition a hazardous waste." (Taylor, A.G., Illinois EPA, "An Overview of Pesticide Disposal Issues in Illinois," 1988.)

Applicators may not read or may not comprehend the label. Commercial applicators who participated in focus group discussions as part of EPA's label utility project volunteered that they rarely read the label. They believed they already knew what it said. Furthermore, they ranked storage and disposal as the least significant label information. (ICF Inc. and SHR Communications & Design, Pesticide Label Criteria and Recommendations: Report on Findings of Focus Group Discussions, 1988.) In a recent Minnesota survey on reasons for triple rinsing, over 70% of respondents said they did so to prevent environmental contamination. Only 20% did so because the label required it. (Hansen, R., Minnesota Department of Agriculture, to T. Bone, EPA-OPP, letters dated October 4 and November 1, 1990).

Labels contain general prohibitions-- e.g., "Do not contaminate water, food or feed by storage or disposal"--that state lead agencies may use to rectify a situation that has caused, or is about to cause, harm, but may not be specific enough to prescribe a detailed standard of conduct. Labels may also contain advisory statements , e.g. "avoid" or "should not," that states find unenforceable. (Kempter, J.,U.S. EPA-OPP Registration Division, to Arty Williams, EPA Registration Division, memorandum dated July 19, 1991.)

Labels may be inconsistent with the toxicity hazards identified in the relevant toxicity studies.(California Senate Office of Research, Regulation in Practice: A Review of the California Department of Food and Agriculture's Pesticide Registration Process, February 1990).

Labels may contradict state policy.

Labels may dictate actions that increase, rather than minimize, waste disposal problems.

Labels may apparently authorize applicators to take action that could violate state law. Virginia and New York have addressed potential conflict between state regulations and labeling. Virginia , by statute, has made it a violation of state law to dispose of containers or their contents contrary to Pesticide Board regulations if those regulations are more stringent than the label. New York, for termite control applications, similarly provides that "in circumstances where the label and these regulations address the same point, the stricter of the label or the regulations must be complied with." (Section 3.1-249.64, Code of Virginia; 6 NYCRR Part 326.2) Most other state pesticide statutes are silent on the issue.

Label Statements on Container Disposal . Through its labeling authority EPA requires that labels contain a statement on container disposal. The labels must contain the language of the guidance documents, PR Notice 83-3 or PR Notice 84-1, or alternative language approved by EPA through amended registration. Many labels incorporate the language of the guidance documents verbatim.

The language suggested by PR Notices 83-3 and 84-1 creates several enforcement dilemmas: First, PR Notice 83-3 implies, for non-household products, that landfilling or incineration are approved state procedures. They are not always approved. Maine, for example, requires return of all restricted use and state limited use containers to designated sites where the containers are inspected and then disposed of. New York prohibits incineration of containers which held volatile herbicides. In addition, many landfills will not accept agricultural chemical pesticide containers. This may be by choice of the owner/operator or by municipal or state permit. Second, triple rinsing mandated by the label for metal, plastic and glass non-household containers, may, in some situations, (TBT paint cans, for example) magnify waste disposal problems by creating additional waste solvents. Third, the procedure for triple rinsing or emptying of bags is not specified. Some states have specified a procedure by regulation or in guidance materials but the procedures vary significantly. Fourth, the household product direction to throw the pesticide in the trash runs counter to the advice and considerable efforts made in many states to institute household hazardous waste collection programs in order to keep toxic substances out of municipal landfills.

Label Statements on Waste Pesticide Disposal. Through PR Notice 83-3, EPA has approved three pesticide waste disposal statements for non-household products. Like the instructions for container disposal, the instructions for waste pesticide disposal also cause enforcement dilemmas. A label following PR Notice 83-3, unless it contains additional instructions, does not specifically address all forms of pesticide waste. It provides no direction for proper disposal or treatment of exterior equipment washes, contaminated clothing or spill clean up debris. Except for those products which PR Notice 83-3 designates "acutely hazardous" or "toxic", it provides no specific direction on disposal of rinsates. Instead, the disposal instruction for non-household products that are neither "acutely hazardous" nor "toxic" is : " Wastes resulting from the use of this product may be disposed of on site." This appears to allow any pesticide wastes to be disposed of by any means , by any type of applicator, at any application site. This may well violate state law. It certainly invites activity likely to cause environmental contamination.

Label Statements on Reuse and Recycling. Labels for products (other than household products) that incorporate the container disposal language of PR Notice 83-3 explicitly authorize certain uses of empty containers:

Metal containers: ...[O]ffer for recycling or reconditioning or puncture and dispose of...
Plastic Containers:...[O]ffer for recycling or reconditioning, or puncture and dispose of...
Glass Containers: [No statement on reuse, recycling, reconditioning]
Fiber Drums with Liners: ...If drum is contaminated and cannot be reused, dispose of...(Manufacturer may replace this phrase with one indicating whether and how fiber drums may be reused.)
Paper and Plastic Bags: [No statement regarding reuse, recycling, reconditioning]
Compressed Gas Cylinders: Return empty cylinder for reuse (or similar wording).
(Emphasis added).

The majority of labels are silent on "reuse" of the container, in contrast to "recycling or reconditioning." This raises the question whether reuse of the container for a purpose other than recycling or reconditioning is allowed. It does take place. A recent article in one agricultural industry magazine suggested making a metal drum into a barbecue pit. The North Dakota Governor's Waste Management Task Force reported in its study, Municipal Waste Management Issues in the State of North Dakota, that "many pesticide containers may have been put to other uses even though this is strictly forbidden." (Cited in "Report of the North Dakota Legislative Council Agriculture Committee," November 1990.) 11% of dealers who responded to the 1990 Montana Department of Agriculture Pesticide Dealer Disposal Survey replied that they used or sold returned or damaged drums and containers for garbage cans.

Some labels explicitly prohibit "reuse" without making any reference to recycling or reconditioning. E.g. Diquat H/A 1 gallon (Valent): "Do not reuse container." The EPA Office of Compliance Monitoring and Office of General Counsel interpret this language to prohibit all possible forms of reuse, including recycling. States do not uniformly interpret labels in accordance with this view.

Further confusion may arise when printed labels affixed to the container may differ in content from reuse statements embossed on the container, leaving the applicator to reconcile the differences. In addition, pursuant to state statutes mandating coding of plastic containers for recycling purposes, pesticide containers may be embossed with a recycling logo although the label states "Do not reuse."

State Disposal Regulations on Open Burning and Burial

Two current means of disposal for agricultural chemical containers are open burning and burial. These practices are both controversial and problematic because it is impossible for state enforcement personnel to determine that the containers are clean prior to disposal. In fact, one of the rationales for container collection programs is that they provide an immediate option to burning or burial. Many labels allow open burning if permitted by state and local authorities. Current labels rarely specifically allow burial, but they do allow disposal by any means approved by state law. Because some states still allow burial, the labels permit this practice to continue.

Open Burning. Open burning, unlike incineration, is the burning of solid wastes in the open, as in an open dump. (40 CFR Part 241.100.) In many places, it offers a convenient way for farmers to dispose of paper bags and plastic containers. Its legal status as a means of disposal for farmers varies among the states: Some states have adopted pesticide regulations that specifically permit burning; Others specifically prohibit it in their pesticide regulations; Some interpret the state air pollution laws to prohibit burning (E.g. Maine, New Jersey, South Carolina, Rhode Island); Some states interpret the air pollution regulations to allow burning; Other states decline to take a public position.(E.g. The Virginia Department of Environmental Protection has not taken a position on the legality of farmers burning containers. Pending an opinion that it is illegal, farmers do burn containers.)

Noting the potential for health hazards, property damage and threats to public safety from unconfined combustion, EPA, under RCRA Subtitle D, prohibits open burning of empty pesticide containers and waste pesticides. (40 CFR Part 257.3-7 and 40 CFR Part 258.24(proposed)) In 40 CFR Part 165.7, however, EPA sanctions the burning of containers in small quantities on farms.

Arizona, Delaware, Florida, Illinois, Iowa, New York, Ohio, Oregon, Pennsylvania and South Dakota specifically permit by regulation open burning of containers under certain conditions, generally limiting the practice to agricultural users burning small quantities of combustible containers on site. Illinois is the only one of these states to authorize burning by dealers and commercial applicators. It does so in its Agrichemicals Facilities Regulations, but permits open burning at commercial agrichemical facilities only until January 1995.

Two states that had for years allowed open burning, North Carolina and West Virginia, no longer do so. North Carolina amended its pesticide regulations as of December 1, 1989, to prohibit the practice.

State pesticide lead agencies that do not address burning directly in the pesticide regulations generally defer to the state air pollution agency for a determination of its legality. For example, in 1986 the Florida Bureau of Air Quality Management submitted a request to EPA to amend its state implementation plan to permit open burning of pesticide containers. Region IV issued a notice of final rulemaking finding that such open burning did not violate the Clean Air Act, although it did violate RCRA Subtitle D, and approved the request. The request has never received final approval by EPA in Washington. The air pollution statutes of a number of states contain a provision similar to the following excerpt from New Mexico's statutes:

open burning is prohibited except for "agricultural management...directly related to the growing and harvesting of crops." This language has been interpreted both to prohibit and to allow burning of containers on site. In California, where the state air pollution statute is similar to that in New Mexico, local Air Resources Boards determine whether open burning is allowed within their jurisdictions. Some allow it. Others prohibit it. Farmers in Nebraska may burn bags. The status of plastic containers is unclear and persons with questions are referred to state air pollution agency personnel. Commercial applicators and dealers may not burn containers. The state has taken enforcement actions against dealers who have done so. Louisiana does not specifically prohibit burning in its pesticide regulations, but the Department of Agriculture plans to amend its regulations to do so. In the meantime, the Department of Agriculture notifies the Department of Environmental Quality of the discovery of any large open burn sites and has sent warning letters to farmers advising them that burning anything other than outer wrapping boxes is illegal. In Indiana, burning is illegal under the state's air pollution laws except pursuant to special permit from the state Department of Environmental Management. The Department will not grant permits and state certification and training programs advise applicants that burning is illegal. (Scott, D., Indiana State Chemist, personal communication, October 1990.)

Even when prohibited by state law, however, as in Maine, open burning of both bags and plastic containers occurs in rural areas because of convenience and state enforcement agencies are not inclined to take action in the absence of practicable recycling or other disposal alternatives. This occurs in New Hampshire, for example. (Cathy Schmitt, New Hampshire Pesticide Control Board, personal communication. The Texas Department of Agriculture in its certification programs instructs growers not to burn containers for their own good. Diane Wilcox, Texas Water Commission, personal communication. Nevertheless, some growers burn plastic containers. Brad Cowen, Texas Extension Service, personal communication, November 1990)

Burial. For many years, burial (the placing under soil cover of pesticide wastes in a site that does not qualify as a sanitary landfill) was a common method of container disposal on farms. Many states still allow the practice, apparently following the guidance in 40 CFR Parts 165.2 and 165.8.

In their pesticide regulations, Arizona, Hawaii, New York, South Dakota and West Virginia specifically authorize burial of containers. Others, like Minnesota, Illinois, New Mexico (bags only), Nebraska, Nevada, North Dakota and Oregon, allow burial of empty containers by farmers on-site pursuant to solid waste regulations.

In those states where burial is legal, it is not a practice which states encourage. Nor is it a disposal method that states can control, once authorized, to ensure that only empty containers are buried or that other pesticide wastes are buried in such a way that groundwater is protected. Burial is still legal in Florida, for example, but Extension brochures advise against it. Buried containers are likely to preclude issuance of a mortgage in Florida when growers try to sell their land.(Dwinell, S., Florida Department of Environmental Regulation, personal communication, November 1990)

Burial is illegal in many states, including: Maine, Texas, Massachusetts, North Carolina, South Carolina.

Exemplary State Programs

Having said that conflicting and overlapping jurisdiction of agencies and statutes characterizes pesticide disposal, that labels are not adequate for their purpose and that many states still allow disposal practices of questionable safety, I do not want to leave you with the impression that all is lost. Many states have adopted regulations or fostered nonregulatory programs to deal with disposal issues that deserve to be emulated by other states and by EPA when EPA writes the package of FIFRA 88 regulations.

Container Collection Programs. Through the organizational efforts of state agencies, seven states- Florida, Idaho, Maine, Minnesota, Mississippi, North Carolina, and Illinois- have conducted or are planning to conduct pilot container collection programs or state wide collection programs. The programs are statutory in Maine, Illinois and Minnesota. The Maine program started in 1985 and remains the only mandatory deposit and return program. In Oregon the agrichemical industry has operated a voluntary program for several years for the collection of metal and plastic containers. A similar voluntary program for plastic containers began in Iowa in 1990. A dealer association in Washington has run a metal container collection program for several years. The National Agricultural Chemicals Association ("NACA") organized collection programs in Vermont, Maine and several other states during 1991 [see paper presented by Dr. Ralph May]. In addition, some registrants, dealers and commercial applicators collect empty containers either by contract or as a service to their customers.

There are three primary goals of the state run and industry organized programs: (1) ensuring that containers are disposed of in an environmentally sound manner, (2) providing a practicable method of disposal for applicators, and (3) recycling containers when feasible, preferably in a closed loop system in which empty plastic pesticide containers are reprocessed into new pesticide containers.

Waste Pesticide Collection Programs. Participants in the 1987 National Conferences and Workshops on Pesticide Waste Disposal concluded that organized free or low cost collection programs are essential if unwanted and unusable pesticides resulting from both agricultural and home use products are to be disposed of safely. Since 1985, more than twenty states have conducted , and several others are preparing to implement, such collection programs for non-household products: California, Connecticut, Hawaii, Idaho, Illinois, Indiana, Kentucky, Louisiana, Maine, Massachusetts, Michigan, Minnesota, Mississippi, New Hampshire, New Jersey, North Carolina, North Dakota, Rhode Island, South Dakota, Texas, Virginia, Washington, Wisconsin, Wyoming. These programs cover pesticides that fall into one or more of the following categories: (1) the pesticide is no longer registered for the purpose for which the holder bought it; (2) the pesticide's physical nature has changed, preventing application; (3) the pesticide's efficacy has been reduced through product deterioration, causing use to be stopped; (4) labels are lost or destroyed , making safe use impossible, and (5) the pesticide, although legally usable, is no longer wanted by the holder. The programs may be statewide or restricted to one or more counties, may be free of charge to participants or may charge a fee, may invite participation of farmers and ranchers only or may include commercial applicators and dealers, may require transportation to a central collection site or operate through pick up on site.

During the conduct of collection programs, states lead agencies frequently receive inquiries about pesticides that are still registered for use but are unwanted. State agency personnel may help to arrange transfer to a person who wants to use the product and is legally able to do so.

Rinsate Collection. States with bulk storage regulations, including Iowa, Illinois, Wisconsin, Minnesota, Vermont, Nebraska (proposed), Indiana, Michigan and Ohio, require as an essential feature of the regulations that all washing and mixing/loading at bulk storage facilities take place within containment areas that capture wastewaters for further application or disposal. Apart from bulk storage regulations, only a few states currently mandate containment at sites used for rinsing, washing, mixing or loading. These states include Wisconsin, Illinois (for lawn care companies), the California Central Valley Regional Water Quality Board, and Oregon.

Suggestions

Based on my review of state programs, I would like to offer several suggestions for issues that should be addressed and ways that practices could be improved to achieve safer pesticide disposal.

Label Improvements. EPA had a vehicle through the moribund Label Improvement Program to address the problems with labels that I mentioned previously. Recently, the Registration Division formed a State Label Issues Committee and began surveying state lead agency personnel to determine ways in which labels may be improved. Through FIFRA 88 regulations, EPA should also address disposal statements on labels. This paper is not an appropriate place to discuss the details of new label language, but I suggest that at a minimum, EPA's efforts should :

(1) Devise a system at EPA, adaptable to states, that enables anyone, agency employee or member of the public, to have ready access to all labels in use for a particular product. These labels should be marked as to date filed by the registrant and date last reviewed by the agency. Among other purposes, this will give interested persons and state agency personnel the opportunity to determine whether the label for products that might have remained in storage for a long time contain the latest statements on proper use or should be superseded in practice by newer language and whether the label warrants review because of its age and intervening changes in pesticide management;

(2) Specifically prohibit burning and burial of containers and any other pesticide wastes;

(3) Remove all language authorizing on-site disposal in favor of references to practices allowed by state and federal law;

(4) Clarify the reuse language on containers so that all containers that may be refilled are marked as such and all containers that may be recycled are marked as recyclable. All other reuse should be expressly prohibited.

Hazardous Waste Transportation and Disposal Requirements. Pursuant to state law and federal law , those who hold waste pesticides may be considered hazardous waste generators subject to RCRA generator requirements and/or operators of hazardous waste storage facilities subject to RCRA storage requirements. This legal knot has made holders reluctant to participate in collection programs, or even to reveal their existence, if the state might use its authority to impose fines for violation of hazardous waste laws. As a result, even anonymous surveys designed to assess the magnitude of waste pesticides underestimate the volume. (Personal communication with state agency personnel in Washington, Maryland, Massachusetts, Maine, Rhode Island and Wisconsin. Elaine Andrews of the University of Wisconsin Extension Service believes based on her experience accompanying farmers on inspections of their facilities that surveys also tend to under-report amounts of unwanted pesticides because holders have simply forgotten what lies hidden in the dark recesses of shelves and barns.) In addition, persons transporting pesticides once they are considered hazardous wastes may need to be licensed hazardous waste transporters. (An analysis of the state regulations affecting the ability of Massachusetts to run a pesticide collection day on the basis that farmers transport their own waste is contained in Memorandum from Peter Bronson, Senior Deputy General Counsel, Massachusetts D.E.P., to Mary Ann Nelson, E.O.E.A., December 14, 1989.) The costs for such transportation may be prohibitive. Each state that has run a waste pesticide collection program has addressed these issues in slightly different ways.

EPA could facilitate proper disposal by adopting a policy that allowed, but did not require, states to determine that unusable and unwanted pesticides destined for a collection program are not waste until they have been brought to a collection site. EPA should also, in all voluntary and involuntary cancellation and suspensions in the future, provide reimbursement for the full cost of transportation. In its September 1991 ethyl parathion cancellation notice, EPA indicated that return to the manufacturer may be arranged for persons who wish to dispose of their stocks. The cost of transportation was not addressed specifically in the notice. Previous state experience with such canceled pesticides as 2,4,5-T and dinoseb, demonstrates that holders will not ship products if they must pay the costs of transportation. Instead, growers hold the canceled products in storage, use, or dispose of them illegally. (Governor John McKernan, Maine, to William Reilly, U.S. EPA, letter dated January 15, 1992)

Date Coding of Containers. Pesticide containers do not generally bear a date of manufacture or an expiration date. EPA ought to require that they do so. Date coding, coupled with some sort of marker to identify whether the product has been held under the labeled storage conditions, would allow holders to use products and holders to exchange unopened pesticides with others who may use them with some assurance of product safety and efficacy. Date coding would encourage rotation of stock on dealer shelves and at the end user site. Date coding would further provide state inspectors with the information they need to determine whether the product label represents the most current information on product management. These consequences of date coding would both reduce the volume of pesticides needing disposal and improve the safety of application.

Cross-contamination. Use in accordance with the label requires, first , that the labeled rates not be exceeded. Most state pesticide agency officials do not view excess loading as a significant impediment to agricultural rinsate use because applicators generally use less than the labeled rate in initial application. Use of rinsates does, however, require careful management. (Taylor, A.G., Illinois EPA, Hanson and Anderson, "Recycling Pesticide Rinsewater," 1986.) Florida recommends limiting rinse water to 5% of the diluent. Minnesota's bulk storage regulations limit the volume of rinsates and sludges to "no more than 5% of any total tank mix for delivery rates of 40g/A or less and 10% for delivery rates of more than 40g/A. Washwater not contaminated with pesticides may be used undiluted." (Minnesota Department of Agriculture, Pesticide Storage Rules, Section 1505.3090(3)(C))

Use in accordance with the label requires, second, that pesticides be applied only to labeled sites and crops. Whenever equipment is used to apply more than one pesticide or when rinsates and washes are collected for later use, the possibility of application to an unlabeled site exists. Few states have grappled with the issue. States and EPA ought to do so. In the Salinas Valley of California, where growers raise many crops on small acreages and use a variety of pesticides, many growers filter container rinsates and reuse the water as washwater for containers and equipment to avoid mixing of rinsates that might result in application of pesticides to an unlabeled site. Recognizing that minimal cross-contamination occurs even in well designed and managed rinsate/washes collection systems, the Wisconsin DATCP has developed a policy that the application of pesticide mixtures, mixed using rinsates containing pesticides not labeled for the intended use site, is not contrary to label directions as long as the concentration of the use-incompatible pesticide is less than 1 ppm in the final spray solution.This policy does not apply to any pesticide found to result in phytotoxicity or illegal residues at these concentration levels. (Wisconsin DATCP, "Pesticide Rinsates, Management and Reuse Guidelines," 1986. Prior to adopting this policy, the Department funded research to assure that major use pesticides present as cross-contaminants did not cause phytotoxicity or residues in treated commodities.)

Sludges from Collection Systems. Rinsates, washes, spills, unused solution and other excess pesticides that cannot be applied in accordance with the label become waste subject to state solid and hazardous waste laws. If states and EPA require, or even promote, the collection of rinsates, washes and spills, they should at the same time address the management of sediments from collection systems. These sludges may present special problems if allowed to accumulate because of their high pesticide content. "[Analyses suggest] that some of the pesticides are concentrating in the sediment, precipitating out, or both. The practical implication is that frequent sediment removal from the mud pit and collection tanks is important to prevent this accumulation, or that agitation of the tank contents prior to withdrawal is necessary to keep these small solids moving through the system."(Taylor, Hanson and Anderson, "Recycling Pesticide Rinsewater," Proceedings of the National Workshop on Pesticide Waste Disposal, EPA 600/9-57/001. U.S. EPA, Cincinnati, Ohio, 1986)

In some states, California for example, any such sludges would likely be required to be disposed of as hazardous waste. In other states, Maine for example, their management would require handling as special waste under rules administered by the state agency in charge of solid and hazardous waste disposal. Illinois requires, in its agrichemical facilities regulations, that all agrichemicals and mixtures that cannot be used in accordance with the label shall be disposed of as special or hazardous waste. (8 Illinois Administrative Code, Section 255.110) The Minnesota bulk regulations are the only containment regulations to refer specifically to sludges. The regulations require that sediments be removed from the trap before it is half full and authorize sludges to be used in tank mix at the same rate as rinsates. Iowa's solid waste disposal rules allow land application of waste pesticides provided that the director of the Department of Natural Resources determines that land application is the best disposal method and the applicant submits and receives approval of a land application plan.

Certification and Training. Certification and training required by state regulation offers one way of disseminating information on proper pesticide management practices, including disposal, and of ensuring that those who apply pesticides have at least some modicum of knowledge in the field. Most states require that commercial applicators, employing either general use or restricted use products, be licensed and, following FIFRA, that non-commercial applicators of restricted use products (private applicators) be licensed. Unfortunately, many persons who may advise others on the use of pesticides or may actually apply them, need not in most states demonstrate competence. These persons include dealers, non-commercial applicators applying general use products, employees of licensed applicators, and homeowners. States could, and some have, adopted regulations that encompass more persons within mandatory certification and training. But one of the most constructive steps lies in the hands of registrants. Dealers are an indispensable source of information on pest and pesticide management for many growers. Registrants have the power, if they choose to use it, to require that dealers who carry their products demonstrate competence in all areas of pesticide management, including appropriate disposal practices.

Even for those persons who do receive state mandated certification and training, printed educational materials often do not reflect the latest information on pesticide management. One of the widely used core manuals, for example, states that burial of rinsates and containers is acceptable. Others may state that burning is acceptable, although the state using the manual prohibits it. States are modifying training material but need both time and money to do so. U.S.D.A. could speed the process through grants programs and through the distribution of state material developed in one state that would be valuable to other states.

Integration of Pesticide Disposal Issues with Recycling and Solid Waste Reduction Legislation and Programs. Pesticide and pesticide container disposal is a subset of the much larger issue of comprehensive waste reduction and recycling. Ideally, pesticide disposal should dovetail with other waste reduction and recycling programs. The following legislation may provide a vehicle for integrating pesticide programs into those efforts or may affect the current programs already underway to minimize pesticide waste.

Thirty three states have enacted comprehensive waste reduction and recycling programs which require detailed statewide recycling plans and/or separation of recyclable and contain one or more other provisions to stimulate recycling: Arkansas, California, Connecticut, Delaware, Florida, Georgia, Hawaii, Illinois, Indiana, Iowa, Louisiana, Maine, Maryland, Massachusetts, Michigan, Minnesota, Missouri, New Hampshire, New Jersey, New Mexico, New York, North Carolina, Ohio, Oklahoma, Oregon, Pennsylvania, Rhode Island, Tennessee, Vermont, Virginia, Washington, West Virginia and Wisconsin. (National Solid Wastes Management Association, Special Report, Recycling in the States: Mid-Year Update 1990. This provides an excellent survey of the status and general content of state recycling legislation.) Most set a goal of 25-50% reduction in solid waste by the mid to late 1990's. Six mandate source separation of certain recyclables and six others mandate municipal separation through means of the community's choosing. Most require development of education materials and information on markets for recyclables and offer grants, loans, or tax credits to stimulate waste reduction and recycling. The 1990 Delaware Recycling and Waste Reduction Act specifically lists "pesticide and insecticide containers" as recyclable material and requires that the Waste Management Authority "shall consider, as part of its source separated recycling and waste reduction program, recovery and use of ...household paint, solvent, pesticide and insecticide containers." (Title 7, Section 6450 et seq., Delaware Code)

Unlike the container collection projects in the agricultural sector that are designed to create a closed loop system, states, municipalities and entrepreneurs have not always tailored the demands of pesticide container collection to recycling programs for municipal solid waste. In some Minnesota communities with curbside collection of recyclables, for example, vehicle operators collect all plastic containers with a neck. This may include pesticides used by homeowners. In other communities, residents may dump these containers in bins at community drop off centers and potentially contaminate the rest of the material. Ideally, pesticide container collection programs in the agricultural sector would be extended to homeowner pesticides and those used in the institutional, lawn care, PCO and other markets that now find their way into the municipal solid waste disposal system.

Packaging legislation to reduce waste and the toxicity of waste may affect the way pesticides are packaged and marketed. Several states have banned packaging that cannot be recycled or given state agencies the power to ban such materials, Iowa and Massachusetts, for example. At the urging of a coalition of environmental groups, eleven state legislatures in 1991 considered model legislation regarding packaging: Maine (L.D. 1371(1991)), Vermont, New Hampshire, Massachusetts, New Jersey, New York, Florida, New Mexico, Illinois, North Carolina and Oregon. The bill, which originated in Massachusetts with the Massachusetts Recycling Initiative, requires all packaging used in a state to meet minimum environmental standards for reusability, recycled material content or recycling. As of mid 1990, eight states had adopted toxicity reduction standards for packaging: Rhode Island, Connecticut, New York, Maine, Vermont, Iowa, New Hampshire and Wisconsin.

Conclusion

Wrestling with pesticide disposal issues successfully is a challenge in which both states and EPA should be willing to engage. It will require changes in long standing disposal practices like burning and burial, recognition by state agencies and EPA that cooperation among agencies and within branches of the same agency is essential, willingness to revise regulations comprehensively and advocate changes in areas like labeling that have been regarded as sacrosanct, and understanding that the symbiotic relationship between the regulatory agency and the regulated community should encourage innovation and at the same time force abandonment of practices that no longer offer the best means of protecting health and the environment.

RECEIVED May 26, 1992

Chapter 3

Managing Pesticide Wastes

Perspective for Developing Countries

Janice King Jensen

Office of Pesticide Programs, U.S. Environmental Protection Agency, 401 M Street S.W., Washington, DC 20460

The issues revolving around managing pesticide wastes in developing countries differ greatly from those in the United States. Among key factors that have caused and continue to aggravate the problem of enormous pesticide waste stockpiles in developing countries are: a) lack of implementable national and regional regulations and educational programs on pesticide waste management, b) inadequate controls over pesticide importation, c) limited technical resources, d) strong demands for empty containers, e) lack of good storage facilities, and f) socio-political influences. Despite these challenges, practical options for the safe disposal of unwanted pesticides are becoming available to developing countries.

The major issues affecting pesticide waste management in developing countries differ greatly from those in the United States. Considerations in the U.S. center on container and rinsate management and disposal regulations. In developing countries, where a single steel drum can cost up to five months' worth of wages, the issues are more fundamental. Too often, developing countries lack the appropriate regulatory and enforcement framework, disposal infrastructure, and qualified personnel needed to manage pesticide wastes. This situation is made all the more difficult by the presence in many developing countries, especially Africa, of vast quantities of now-obsolete pesticides, many of which were donated for controlling migratory pests.

Developing countries face a number of specific problems in tackling pesticide wastes (*1*).

The views expressed in this paper are those of the author, and do not necessarily represent policies of the U.S. EPA.

Poor Control Over Pollution and Waste Disposal

A common complaint heard in the U.S. is that there are too many authorities and regulations governing pesticide waste disposal. This contrasts sharply with the situation in most developing countries -- especially Africa -- where pesticide regulations and their enforcement are essentially nonexistent.

With little regulatory framework, a developing country has little control over the type, quality and quantity of imported pesticides. In addition, the lack of a regulatory framework, especially the lack of adequate enforcement capability, makes it difficult to implement worker safety measures and assure proper use and disposal of pesticides.

Countries without effective legislative controls may suffer the consequences of poor and even fraudulent practices, as occurred during the production of the 1979 Kenya coffee crop. Coffee, Kenya's largest single earner of foreign exchange, was threatened because of adulterated pesticides brought in to control coffee berry disease (*2*). The formulation contained only 45% difolatan, not 80% as specified on the label, and had been cut with chalk. The Kenya Ministry of Agriculture attributed a 7% - 8% drop in coffee production that season to the substandard pesticide. This incident was a catalyst for the Kenyan government to establish the Kenyan Pest Control Products Act in 1982 (*3*).

A similar problem occurred in Cameroon in 1984 (*4*). Paraquat, with an active ingredient content of 2.4%, rather than the 24% stated on the label, was applied by unsuspecting subsistence cocoa and banana farmers. Because of incidents like these, fraudulent practices involving pesticides have become a problem of paramount concern to officials in developing countries (*5*).

Problems are by no means limited to isolated shipments of large quantities of pesticides. Unfortunately, poor enforcement is a ubiquitous problem in developing countries in Asia and Africa (*6-7*). For instance, Thailand has a regulatory scheme and analytical laboratories. However, about 50% of the pesticides in its marketplace are substandard, according to criteria set out by the Food and Agriculture Organization (FAO) of the United Nations (*8*).

Labelling problems add significantly to poor control of pesticides in developing countries. Typical problems include:

- labels made of low-grade paper that disintegrate in sunlight;
- labels printed similar in color to the drum, and are therefore difficult to read;
- labels printed in English, German, or Japanese, but not in the local language; and
- no labels at all.

Inadequate storage also is an obstacle. Typical storage problems include:

- pesticides being stored in the sun under severe tropical conditions for long periods of time;
- pesticides being stored in buildings not designed to hold pesticides, typically with inadequate ventilation;
- enhanced degradation of pesticides and their containers because of poor storage conditions;
- increased environmental contamination at storage sites due to failed containers leaking pesticides onto the floor, commonly dirt;
- increased human exposure, especially by inhalation, at storage areas where containers are leaking; and
- inadequate pesticide warehouse management.

These are the day-to-day problems that add significantly to the ever-growing pesticide waste problems in developing countries.

Lack of Awareness of the Hazards of Pesticides

In many developing countries, there is a general lack of awareness of the hazards associated with pesticide use. Inadequate education and personnel training are obstacles to developing awareness; at the pesticide user level, illiteracy is the norm. Many problems associated with pesticide misuse could be eliminated if the end user could read and follow the information, or pictograms, on the label. This would solve many problems in the United States, as well.

The following are examples of typical misuse problems in developing countries:

- reuse of "empty" pesticide containers for water, food and grain;
- improper storage of highly toxic pesticides, often in soft drink or beer bottles;
- increased applicator exposure because spray personnel often do not wear the protective clothing advised on the label, either because the protective clothing is not available, or it is too uncomfortable to wear in the heat;
- improper dilution of pesticides, thereby under- or overdosing crops; and
- use of a pesticide on a crop not on the label.

Rinsate management is a topic of considerable interest in the United States and Europe. However, in most developing countries, rinsate management is not even considered a problem. Perhaps one reason for this is that the farmers are unaware of the hazards associated with rinsate contamination. Another reason may be that the small-scale farmer primarily uses a backpack sprayer, and there is less rinsate residue with this type of small application equipment.

Applicator training courses in developing countries discuss rinsate management, but do not focus on it. The main point in most courses is to ensure that rinsates, whether excess from tank mixes or water from cleaning equipment, are not put into streams or other water sources. There is no regulatory framework like in the United States to dictate which rinsates are wastes and how they should be treated.

Stockpiles of Pesticides Awaiting Disposal

In many African countries, pesticide donations comprise perhaps 80% of all pesticide imports (*9*); in some countries, they comprise 100% (*10*). Although donor organizations contribute these pesticides with good intentions, their gifts can cause more problems than they solve: Most of the large stockpiles of pesticides awaiting disposal can be directly linked to these donations (*11*). Inadequate storage of these stockpiles may make a bad situation worse, with increased numbers of deteriorating containers.

Records show that some donors give pesticides far in excess of a country's requirements. Many of these excessive pesticide shipments are linked with other commodities in an aid package (*12*). For instance, a country may be able to receive highly desired commodities, such as Toyota vehicles, only if the aid package consists of pesticides of equal value, regardless of whether a legitimate need for the pesticides exists.

This situation in Benin is so serious that the German technical aid agency GTZ refuses to assist in pesticide disposal problems until the Benin government begins rejecting unneeded pesticides (*13*). In Guinea-Bissau, another recipient of a Toyota-linked aid package, the stocks of donated dimethoate and fenitrothion far exceed projected demands. Because storage under tropical conditions shortens the shelf-life of those pesticides, disposal problems are inevitable (*10-11*).

Some developing countries receive large quantities of pesticides that they have not requested and do not need. For instance, at the same time Somalia purchased cumachlor in vast quantities for a rodent outbreak in 1978, a large, unsolicited donation of rodenticides also was provided (*14*). Somalia now faces a major problem disposing of its excess rodenticides.

Stockpiles of pesticide for disposal in some developing countries can be traced back to substandard products, such as the paraquat problem in West Africa in 1984, mentioned above (*4*). In addition, obsolete pesticides have become a major problem, representing the bulk of the current pesticide disposal problem in many developing countries. Chemicals such as dieldrin and BHC, donated in the 1960s, are now considered by the international community to be obsolete. In a recent collaborative effort involving USAID, GTZ and Shell Chemical, about 50 metric tons of dieldrin were collected in Niger and shipped to Europe for disposal. This volume is typical for most countries in the Sahel region (*15*).

Table I provides an overview of the magnitude of the pesticide disposal problem in Africa. This is a low estimate of the disposal problem. For every disposal site, there are usually contaminated soils, drums and solvents requiring disposal.

Table I. Obsolete Pesticides for Disposal in Africa

Country	Metric tons
Algeria	898+ (*16*)
Angola	50 (*12*)
Benin	+ (*13*)
Burkina Faso	93 (*17*)
Botswana	18 (*12*)
Cape Vert	23 (*17*)
Chad	114 (*17*)
Ethiopia	440+ (*18*)
Gambia	85 (*17*)
Ghana	40 (*17*)
Guinea-Bissau	12 (*10*)
Ivory Coast	3+ (*17*)
Kenya	48+ (*18*)
Libya	300 (*16*)
Madagascar	+ (*13*)
Malawi	75+ (*12*)
Mali	141 (*17*)
Mauritania	380 (*17*)
Mozambique	+ (*12*)
Morocco	2339 (*17*)
Niger	+[1] (*17*)
Namibia	+ (*12*)
Senegal	131+ (*17*)
Somalia	103+ (*18*)
Sudan	1080 (*18*)
Tanzania	+ (*12*)
Tunisia	500 (*16*)
Zambia	85+ (*12*)
	6958 metric tons

+ pesticide stocks for disposal identified, not quantified

1 50 metric tons removed 5/91

Figure 1 provides a graphic representation of the magnitude of the disposal problem in Africa.

Limited Resources

Insufficient infrastructure, repackaging facilities, storage facilities and training all are obstacles to effective management of pesticide wastes in developing countries.

Although expensive, the infrastructure needed to dispose safely of large quantities of pesticides is available in Europe and the United States. In developing

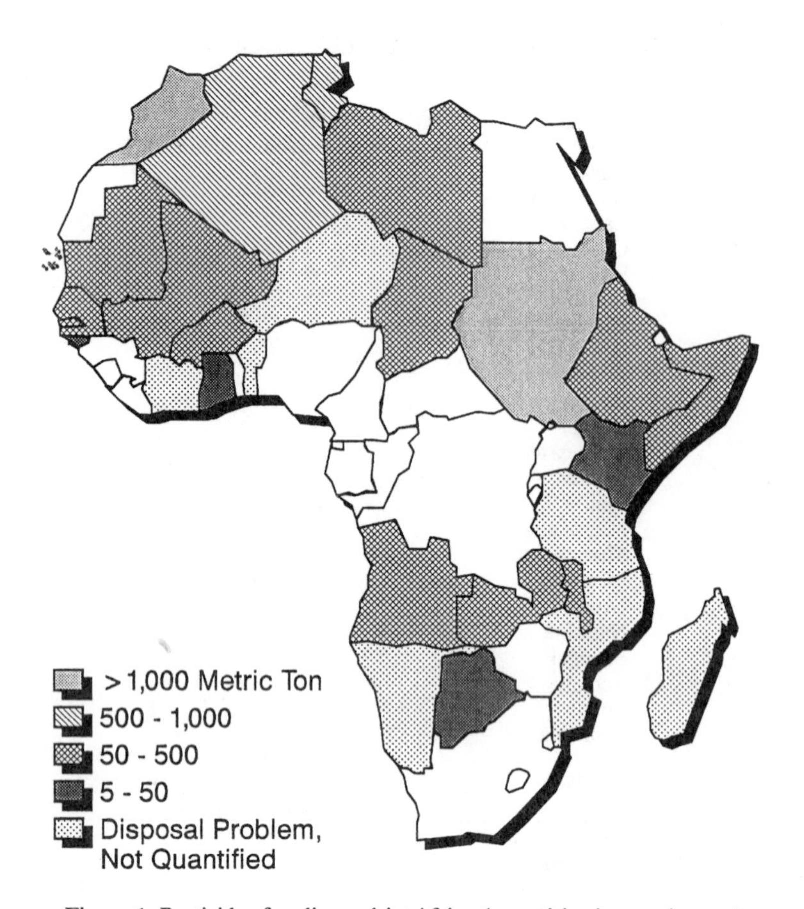

Figure 1. Pesticides for disposal in Africa (quantities in metric tons).

countries, where organochlorines represent the bulk of pesticides requiring disposal, high-temperature incineration is the only viable disposal technology. Yet not a single high-temperature hazardous waste incinerator exists on the entire African continent. Such a lack of infrastructure severely limits the options available for safely disposing of large quantities of pesticides.

Sometimes, government officials do not even have a way of knowing what stocks are in-country, the condition of these stocks, or when they were imported. This lack of infrastructure for inventorying -- unfortunately, a common problem -- can lead to ordering more pesticides than are actually needed, even when viable stocks are already in-country.

Some developing countries also lack pesticide repackaging facilities, so large quantities of pesticides imported in 200-liter drums may go unused. Eventually, they become a disposal problem. In one effort to solve such a problem, Somalia negotiated to obtain 5,000 liters of ULV fenitrothion in 5-liter containers, rather than 10,000 liters in 200-liter drums (*14*).

The shortage of adequately trained personnel at all levels of government is a major obstacle to establishing and maintaining waste management programs in developing countries (*1*).

The Economics of Container Management

Container management in both the United States and developing countries is often driven by economic considerations. Current questions in the United States include: How do we minimize the number of containers needing disposal, since landfill operators often will not accept them? What is the best way to dispose of containers, recycle their materials or recover their energy value? How soon can we start using small-volume containers that can be refilled at the dealer level? Will EPA enforce against us if we only do two quick rinses? Why can't I burn my jugs and bags on my own land?

In developing countries, the major questions on container management include: How can we ensure that empty containers are not misused, since they are such valuable commodities? How can we get pesticides supplied in smaller containers for small-scale end users, since we have no facilities to repackage pesticides safely? How can we make sure that our containers will have labels we can read and will hold up under extreme tropical storage conditions?

The differences in viewpoints revolve primarily around the different economies of the United States and a developing country, not on potential hazards.

The economics of the poorer countries, where pesticide containers are valuable commodities, favor reuse. Empty pesticide containers are used for water, food, grain and fuel storage. These are some of the realities: In Guinea-Bissau, West Africa, a new 200-liter steel drum costs the equivalent of about US $50 - $100 in the local market, three to five times the monthly salary of a typical semi-skilled worker (*10*). In Sudan, a 200-liter drum costs the equivalent of a week's wages of a driver in Khartoum (*19*).

Here, most pesticide formulations and containers are designed around an expected two-year stay in the channels of trade. But in developing countries, lengthy ordering procedures and shipping times mean pesticides may be in storage, usually under temperature extremes, for far longer than two years.

Officials in developing countries are starting to realize that they can require all pesticides destined for their country to be formulated for use under tropical conditions, provided in long-life containers, and marked with extra-sturdy labels that will hold up under extremes of weather. The international pesticide industry agrees that these are good practices to help avoid future disposal problems (*20*).

Socio-political Factors

Developing countries may focus on other real and urgent problems and not see pesticide waste management as a pressing need or immediate political goal (*1*). At a 1985 regional workshop on pesticide legislation in Togo, 10 African countries concurred that the lack of legislation mainly stemmed from political, not technical, causes (*5*).

Without public awareness of the dangers of improper reuse of "empty" pesticide containers, improper disposal of pesticides and other hazardous wastes, too often there is insufficient public demand for action.

Viable Disposal Options

Options currently recommended by the International Group of National Associations of Manufacturers of Agrochemical Products (*20*) for the disposal of unwanted pesticide stocks include: 1) high-temperature incineration using a small-scale fixed incinerator; 2) large-scale fixed incinerator; 3) mobile incinerator; 4) cement kiln incinerator; 5) chemical treatment; and 6) long-term storage. Co-firing pesticides such as the organochlorines as a co-fuel in a cement kiln is an efficient method of disposal that has been successfully carried out in Pakistan (*21-22*).

Cement kiln incineration holds the most promise in developing countries for in-country disposal of large quantities of pesticides. However, because stack gases cannot be adequately controlled in some kilns, cement kilns that are old and inefficient should not be considered.

Cement kiln incineration is a relatively inexpensive, efficient way to destroy certain pesticides, especially organochlorines. However, in developing countries, there are political considerations that also may be a hindrance to this disposal method. In Sudan, for example, all cement kilns were government-run by the Ministry of Industry, whereas the pesticide wastes earmarked for disposal belonged to the Ministry of Agriculture (*23*). For the kiln manager, who is concerned about the quality of his cement, there are few economic incentives to use his kiln for pesticide disposal. Therefore, cement manufacturers need to be convinced that co-firing pesticides does not adversely affect the quality of their product (*20*).

Mobile incinerators are being developed that may be appropriate for future use in developing countries. But for now, they represent an expensive option. In Pakistan, where there were an estimated 5,000 metric tons of pesticides for disposal, the estimated cost to use a mobile incinerator to destroy those pesticides was $17.5 million (*22, 24*).

Limitations in portability also may be a problem in Africa, where most of the pesticides requiring disposal are in remote, sometimes roadless locations. Such incinerators also can introduce harmful emissions if not carefully controlled and operated (22).

Another option was efficiently carried out recently in Niger as a collaborative effort between its government, Shell International Chemical Company, the United States Agency for International Development, and the GTZ. The operation collected and repackaged the pesticide wastes, and transported them to a developed country (the Netherlands) for incineration at a dedicated hazardous waste facility (*15*).

Conclusions

Pesticide waste management issues in the United States differ greatly from those in developing countries. The dominant domestic issues are container and rinsate management and regulatory constraints on disposal. In developing countries, the issues are more fundamental and often are determined by economics. For instance, containers are so valuable, their reuse is common. Also, because of competing priorities and limited resources, most developing countries lack regulations governing pesticide management, leaving those countries vulnerable to inappropriate and fraudulent practices.

A key problem in the poorer countries is pesticide donations -- of inappropriate types; in poor packaging in inadequate storage facilities; or of large, unusable quantities. Also, the lack of a disposal infrastructure limits the options of poor countries for safely disposing of large quantities of pesticides. As a first step to avoiding these problems, donor organizations and recipient countries should increase coordination on pesticide-specific needs.

To help avoid future disposal problems, officials in developing countries should continue to require all pesticides destined for their countries to be formulated for use under tropical conditions, provided in long-life containers and marked with long-life labels. Pesticides should not be provided if there is inadequate storage for the pesticides.

In the meantime, practical options for disposal should focus on high-temperature incineration of the organochlorines, either in-country or by transport to a developed country for proper disposal.

Literature Cited

1. World Bank. *The Safe Disposal of Hazardous Wastes, the Special Needs and Problems of Developing Countries*, World Bank Technical Paper Number 93, Batstone, R., Smith, Jr., J. E., Wilson, D., eds., 1989.
2. Geschwindt, S. "Africa: Where the West Dumps Its Poisons," *New African*, September 1981.
3. Kibata, G. B. In proceedings from a regional workshop in Nairobi, Kenya, sponsored by REDSO/ESA, USAID, "Constraints in the Implementation of Kenya's Pesticide Legislation," Jensen, J. K., Stroud, A., Mukanyange, J., eds., 1985.
4. Deuse, J. "Contrefacon et Falcification des Produits Phytosanitaires: un Fleau pour les Pays en Voie de Developpement," presented at the conference Journee d'etude en Phytopharmacie, Dec. 12, 1984.
5. Weiler, E. The West African Regional Workshop on Pesticide Legislation, presented at the American Chemical Society meeting, New York, New York, 1986.
6. Asian Development Bank. *Handbook on the Use of Pesticides in the Asia-Pacific Region*, 1987.
7. FAO. "Report of the Sub-Regional Workshop on Pesticide Management for Western Africa," 1989.
8. Jensen, J. K. "The Pesticide Situation in Thailand," prepared for USAID/Thailand, 1986.
9. Bryant, M. "Agrochemical Market Hinges on Aid Priorities," *African Economic Digest*, Aug. 31, 1984.
10. Jensen, J. K. Trip report, "Pesticide Storage and Disposal in Guinea-Bissau, West Africa," prepared for the USAID Guinea-Bissau Food Crop Protection Project III and the Government of Guinea-Bissau, 1990.
11. Jensen, J. K. "Pesticide Donations: The Need for Better International Collaboration," presented at the International Conference on Pesticide Disposal, 1990.
12. FAO. "Draft Report on the State of Pesticide Management in the SADCC Sub-Region for the FAO/SADCC Subregional Workshop on Pesticide Management, Harare, Zimbabwe," prepared by H. van der Wulp, 1991.
13. Schimpf, W., Deutsche Gellschaft fuer Technishe Zusammenarbeit (GTZ), personal communication, May 6, 1991.
14. Noor, B. F., Director, Crop Protection, Ministry of Agriculture, Somalia, personal communication, June 16, 1987.

15. Jensen, J. K. Trip Report, "Pesticide Disposal, Niger, West Africa," 1991.
16. FAO. "Pesticide Stocks in Algeria, Morocco, and Mauritania," prepared by Geoff Jackson, 1990.
17. USAID. Country reports presented at the USAID Pesticide Disposal Conference, Niamey, Niger, Jan. 21-26, 1990.
18. World Environment Center. "Evaluation of Disposal Options Re: Pesticide Waste in East Africa -- the Sudan, Ethiopia, Kenya, and Somalia," sponsored by USAID/OFDA, 1987.
19. World Bank. "Recycling Pesticide Containers (in Sudan)," prepared by Hunting Technical Services Limited, Boundary Way, Herts HP2 7SR, England, 1989.
20. GIFAP. *Disposal of Unwanted Pesticide Stocks*, 1991.
21. Huden, G. H. "Pesticide Disposal in a Cement Kiln in Pakistan, A Pilot Project, OFDA/USAID," 1988.
22. Huden, G. H. "Pesticide Disposal in a Cement Kiln in Pakistan, Report of Pilot Project, and Test Results from a Pilot Burn of Overage Pesticides, OFDA/USAID," 1990.
23. Jensen, J. K. Trip Report, USAID/OFDA Pesticide Disposal Survey Team to Sudan, Ethiopia, Kenya and Somalia, May 28-June 25, 1987.
24. USAID. "Pesticide Disposal in a Cement Kiln in Pakistan, Report of a Pilot Project," presented by Gudrun Huden, and a report by D. G. Khan, J. Chehaske and H. Yoest, "Test Results from a Pilot Burn of Overage Pesticides," sponsored by USAID/OFDA, 1990.

RECEIVED February 3, 1992

CONTAINERS

Chapter 4

Container Minimization and Reuse

Scott W. Allison

Monsanto Agricultural Company, 800 North Lindbergh Boulevard, St. Louis, MO 63167

The agricultural chemicals industry shares with its customers, and with state and Federal regulators, a deep concern about the pesticide container disposal problem. Since 1984, a task force of the National Agricultural Chemicals Association (NACA)--the industry's trade association--has been working aggressively to devise and promote solutions to this problem. The NACA Container Management Task Force shares with the EPA a conviction that the best way to attack the problem is by source reduction--by minimizing the burden placed on the environment by the use of single trip, non-refillable, containers for our products. Indeed, source reduction resides atop NACA's container management program hierarchy (Figure 1).

There is no single "best" approach to source reduction. Thus the industry is exploring several complementary avenues.

Lightweight Containers. Perhaps the most obvious approach is the "light-weighting" of existing containers; for example, reducing the amount of plastic used in jugs, the paper in cartons, etc.. This approach doesn't decrease the number of containers that must be dealt with but it does reduce the total burden placed on landfills and other disposal options. It occurs almost "naturally" as manufacturers strive to control their packaging costs. Unfortunately, it results in only incremental reductions in the container disposal problem---it's clearly evolutionary not revolutionary.

Product Improvements. Another, less obvious and far more technically challenging strategy, is the development of pesticides with increased efficacy or activity. When a smaller quantity of pesticide product is required to do a particular job, a smaller amount of packaging material is likely to be required as well. The industry has discovered several entirely new chemical families in the past few years the are very efficient pesticides. A variety of new commercial products based on these chemistries have been introduced. Examples that come to mind are the sulfonylureas (eg Classic and Pinnacle produced by DuPont and

0097–6156/92/0510–0030$06.00/0

Beacon by Ciba-Geigy) and the imazaquins (Scepter from American Cyanamid). These products, which are applied at rates as low as a few grams per acre, are replacing older products which are used at pounds per acre rates. They are packaged in smaller containers and container disposal problems are thereby reduced.

Changes in product formulation can also have a beneficial impact on container minimization. For instance, Monsanto's recent announcement of the development of a new dry tablet formulation for its Roundup herbicide implies a significant decrease in the number of plastic bottles, trigger sprayers and cartons that it will use to meet the needs of its lawn and garden customers.

New Packaging Technology. The introduction of containers made from new materials can also help alleviate the container disposal problem. For instance, the use of water soluble (polyvinyl alcohol) bags for packaging dry powder or granular products is being pursued by several ag chem producers (Figure 2). Soluble pouches containing the product are added by the user directly to the spray or mix tank. There they dissolve and release their contents. In essence the primary package disappears and no disposal problem remains. It is clear that this type of package reduces the amount of potentially hazardous pesticide contaminated packaging. This is a step in the right direction. It is not so clear that water soluble packaging results in a reduction in the total amount of packaging material used and thus qualifies as source reduction. Water soluble bags require a substantial amount of secondary packaging--moisture resistant foil pouches, cartons and shipping cases--in order to survive the rigors of transportation and storage prior to use. This packaging material remains to be disposed of. Thus the environmental burden from the total packaging system may not be reduced compared to that created by other forms of packaging. Still, it's too early to write off this approach. It certainly deserves continued development.

Reusable Containers. During the 80's, the industry made very dramatic progress in reducing its single trip container usage through yet another strategy--the use of reusable, or refillable, containers. The remainder of this paper will focus on the remarkable shift that has occurred in the way that the agricultural chemical industry delivers its products to its customers and address developments, now on the horizon, that will further decrease the use of non-refillable, single trip containers.

History. The first significant use of returnable/reusable containers for pesticides began in the Mid-West--the American corn belt--in the early 70's. In this part of the country, conditions are optimal for the use of bulk delivery systems for liquid pesticides. Farming practices in the region are the key. Agriculture is, of course, dedicated primarily to the production of corn and soybeans. Farm acreages are large and herbicides are used in substantial quantities to suppress weed growth in these crops. Annual consumption of herbicides by a typical mid-western corn or bean farmer often amounts to

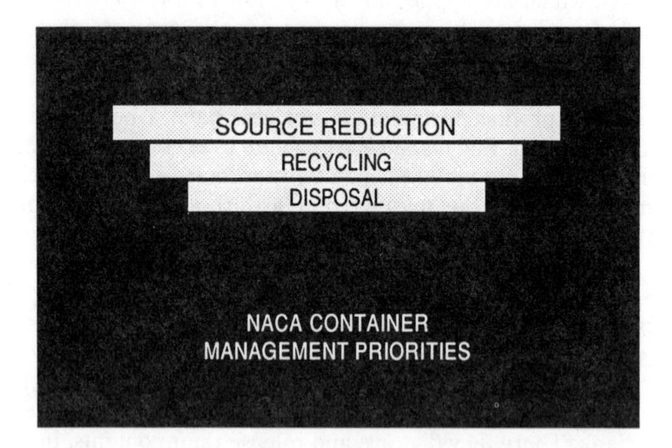

Figure 1. Container Management Program Hierarchy

Figure 2. Water Soluble Packaging

hundreds of gallons. The practical problems associated with simply opening and emptying hundreds of small containers, to say nothing of managing their disposal after use, are substantial. Thus herbicide delivery in large returnable containers has a natural fit.

Logistical Issues. From the chemical producer's and retailer's point of view the shift to bulk packaging was a major challenge. Traditionally, agricultural chemicals have been manufactured in large centralized facilities where the successive steps of: a) synthesizing the active chemical ingredient, b) combining the active with solvents, surfactants, etc. to create a useful formulation and c) packaging the finished product, occur. The packaged product is then shipped, often through a complex distributor/dealer network, to the farmer. Conceptually, at least, bulk delivery is much simpler. The synthesis and formulation steps remain the same, but packaging is omitted and the product is shipped in bulk tanker trucks directly from the production site to the local ag chem dealer, who transfers it to the farmer during the use season.

Many logistical problems had to be overcome before bulk delivery of pesticides could grow. First, the manufacturers had to modify their facilities to permit the loading of tanker trucks. They also had to invest in large capacity tanks for pre-season storage. The retailers also had a problem because they didn't have sufficient tankage at their locations to accommodate the delivery of the various products that were offered. These facilities all had to be put in place before the use of bulk pesticides could begin. This took time and the investment of significant sums of money by the industry and by the ag chemical distributors and dealers.

Container Development. Of course, containers also had to be developed to move the products from the dealer's site to the farmer's field. That's where reusable containers--commonly known as minibulk tanks--came in. The first minibulks were fabricated stainless steel, or molded plastic, tanks which had originally been intended for industrial uses (Figures 3, 4). They served their purpose but they lacked many features that farmers needed. None came equipped with pumps and meters to facilitate easy, accurate measurement of the product into the farmer's spray tank. Many were not designed to tolerate the physical and environmental stresses that are common in agricultural service.

In 1986, the first fully integrated minibulk system was introduced by Monsanto Agricultural Company (Figure 5). It incorporated a built-in pumping/metering system specifically designed for herbicide use and a pallet to facilitate handling. Other producers quickly introduced their own tank designs and a race to develop the "best" minibulk system began (Figures 6, 7, 8). Tank design has become an important marketing tool. Most of today's minibulk tanks have capacities in the 100-200 gallon range, although tanks as small as 60 gallons are now in service. All have integral pumping/metering systems which are not only convenient but also help to reduce worker exposure during use.

Performance Specifications. The industry recognized that the use of minibulk tanks involved important tradeoffs. Specifically while the shift to large reusable containers improved user convenience and reduced disposal problems,

Figure 3. Generic Stainless Steel Minibulk Tanks

Figure 4. Generic Plastic Minibulk Tank

Figure 5. First Generation Integrated Minibulk System

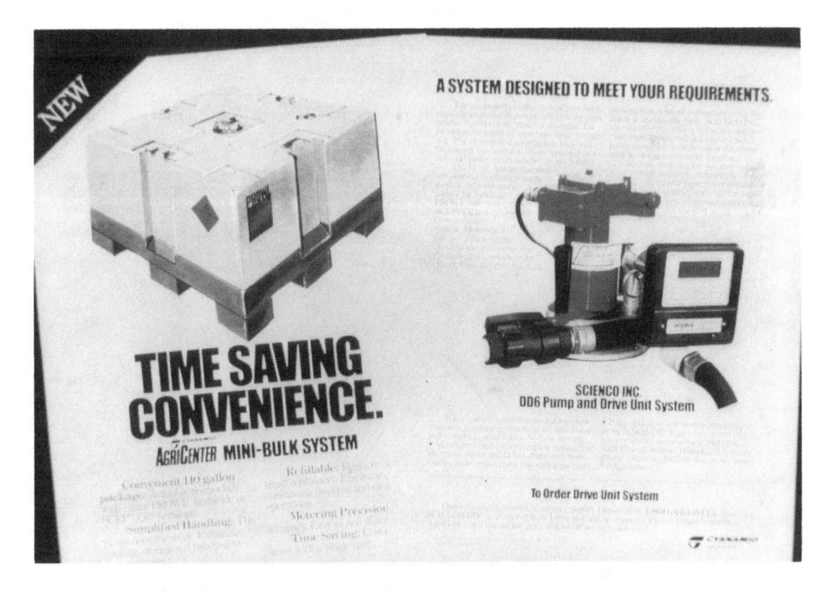

Figure 6. Modern Integrated Minibulk System

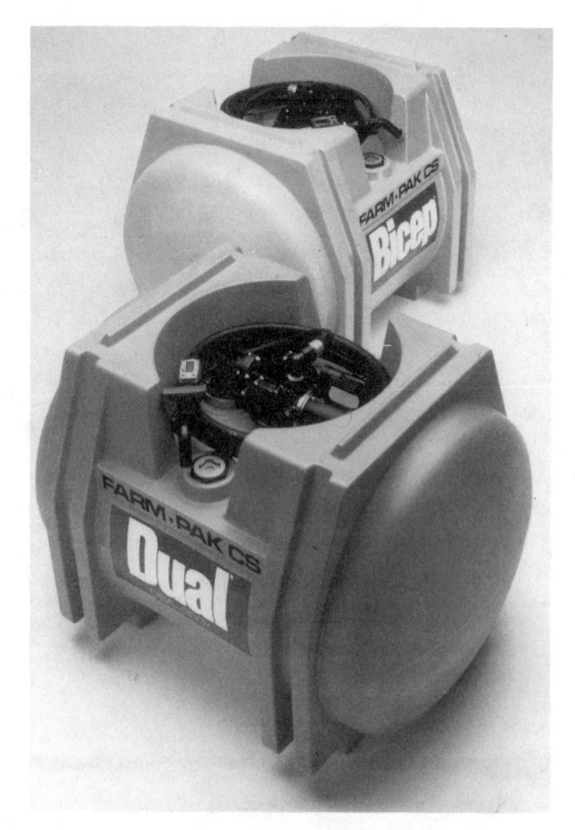

Figure 7. Modern Integrated Minibulk System

Figure 8. Modern Integrated Minibulk System

it increased the risk of environmental problems if tanks failed in service. The answer to this problem was the creation and adoption of a set of detailed design/performance standards for minibulk tanks. These standards were developed by the Midwest Agricultural Chemicals Association (MACA) and are commonly referred to as the "MACA-75" guidelines.

The MACA guidelines were introduced in 1987. They incorporate many elements of existing DOT standards for bulk shipping containers but they also include many features which are specific to ag chemical tanks. They cover both general design considerations--eg. materials of construction, opening/closure design and handling characteristics--and specifications for performance in vibration, drop and hydrostatic pressure tests. Production quality control testing requirements are spelled out in the standards, as are marking, labelling, in-service filling, maintenance and inspection procedures. The MACA-75 guidelines are now being revised and upgraded to reflect experience gained since they were introduced. Among the new items that will be addressed in the revised guidelines are improved protection for external "appurtenances" such as pumps, meters, stacking performance tests, and limitations on permissible service life. The updated guidelines are expected to issue in 1992. Since the MACA-75 standards were issued, they have gained general acceptance by the industry and new tanks are being designed with them clearly in mind.

New Developments In Reusable Containers. In the late eighties several other developments took place which have increased the use of returnable/reusable containers.

First, the use of mini-bulk tanks has begun to spread outside the Mid-West. The efficiency and environmental advantages of reusable containers have been recognized by others. The logistical infrastructure to support bulk deliveries is being expanded and minibulk tanks are now in service in the wheat fields of the Dakota's, the citrus groves of Florida and the vineyards of California. They are also being introduced in non-crop applications like forestry and vegetation control along railroad, highway and utility rights of way.

Second, a new family of reusable containers for liquid products has been introduced--the so-called small volume returnables or "SVR's". Minibulks are generally defined as tanks in the 60 to 600 gallon size range. SVR's fill the gap below 60 gallons. Most are in the 15 to 30 gallon range. SVR's are intended for situations where chemical usage is too small to make the larger tanks a realistic alternative to single trip containers. The first SVR's (Figure 9) were stainless steel containers which bore a striking resemblance to beer kegs. These containers are relatively expensive and several companies, Monsanto and Rhone-Poulenc included, are now testing lower cost plastic SVR's (Figures 10, 11).

SVR's are generally being returned for refilling by the manufacturer rather than by the retail dealers. One reason for this difference from the normal practice for minibulks is an EPA policy, commonly known as the "56 Gallon Rule", which prohibits bulk dealers from filling containers smaller than 56 gallons. The rationale for this policy--which was instituted by the Agency in 1976--is a legitimate concern about the filling of potentially unsafe containers by

Figure 9. Stainless Steel Small Volume Returnable Container

Figure 10. Plastic SVR Container

the retailers. Unfortunately the policy is in direct conflict with the industry's and the Agency's goal of reducing the use of single trip containers. Strict compliance with the policy, in fact, encourages continued use of single trip containers. Modification of this rule is now being considered by the Agency. Removal of this roadblock will encourage the use of SVR's.

Finally, the use of returnable containers is now being extended to dry products. Several different avenues are being explored. American Cyanamid has developed, in conjunction with the equipment manufacturer, John Deere, a container that is essentially an SVR for their granular insecticides (Figures 12, 13). These containers, which hold about 30 pounds of product, couple directly to the hoppers on Deere's planters. They offer the additional advantage of minimal potential for worker exposure to the product. Once emptied, the containers are returned to the manufacturer for refilling. Monsanto has pioneered the use of 1000 lb. "bulk bags" or "super sacks" for use with its granular herbicides (Figure 14). These bags, which have previously been used for shipping industrial commodities, are returned to the plant for refilling after use. The company is also exploring the delivery of bulk quantities of granular products directly to the dealer. ICI is reportedly developing a "granular minibulk" tank which resembles a liquid minibulk tank but which replaces the pump with a blower system to transport the granules.

Development work on reusable containers for dry products is not as advanced as that for liquid containers. It is clear, however, that this is an area that will receive increased attention in the future, especially given the EPA's concern about the disposal of the multiwall bags that are most often used for granular products.

Progress On Container Minimization. All of this effort has begun to result in measurable progress in reducing the number of one-way containers that are being used by the ag chem industry. Until 1988, when NACA began to gather information from its members, no reliable data on pesticide container usage were available. Data collected by NACA since then show a clear reduction in the number of plastic and steel single trip containers used for liquid products (Figure 15). The data, which are presented in terms of the number of gallons of product packaged in each type of container, clearly shows a reduced reliance on single trip containers. In 1991 the industry's consumption of plastic jugs was almost 25%, or about 12M units, lower than in 1988. Steel and plastic pail usage decreased about the same amount in percentage terms during the period. Steel and plastic drum use has shown a smaller decrease, but it should be noted that many drums are, in fact, being returned, reconditioned and reused. It's unfortunate that data don't exist for the earlier part of the decade, because it would make the industry's accomplishments in source reduction even more dramatic. Monsanto's experience in the growth of bulk product shipments demonstrates this point. The company's volume of bulk shipments increased five-fold during the 80's. Its reliance on one way containers has been decreased proportionately.

Figure 11. Plastic SVR Container

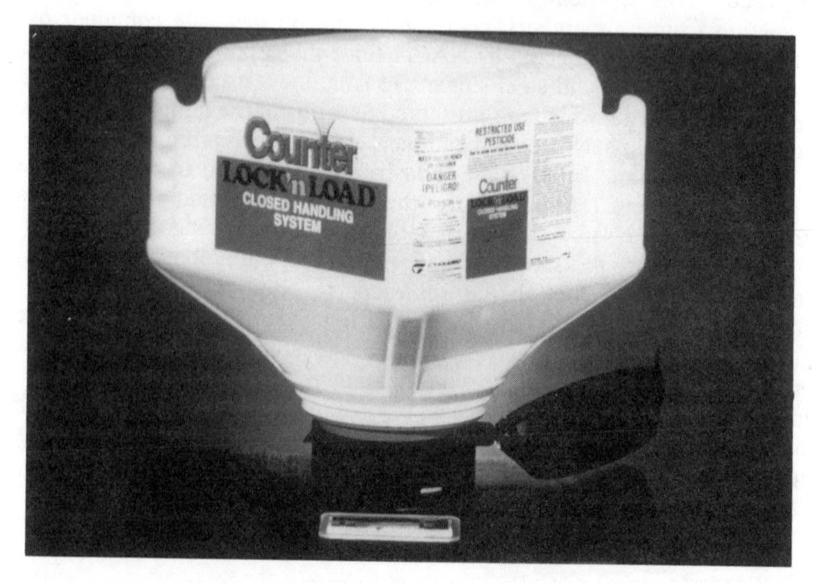

Figure 12. SVR For Granular Products

Figure 13. SVR For Granular Products

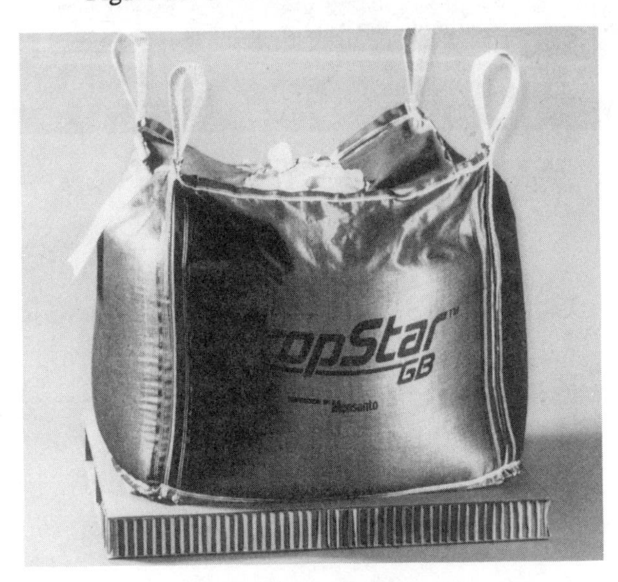

Figure 14. Bulk Bag For Granular Products

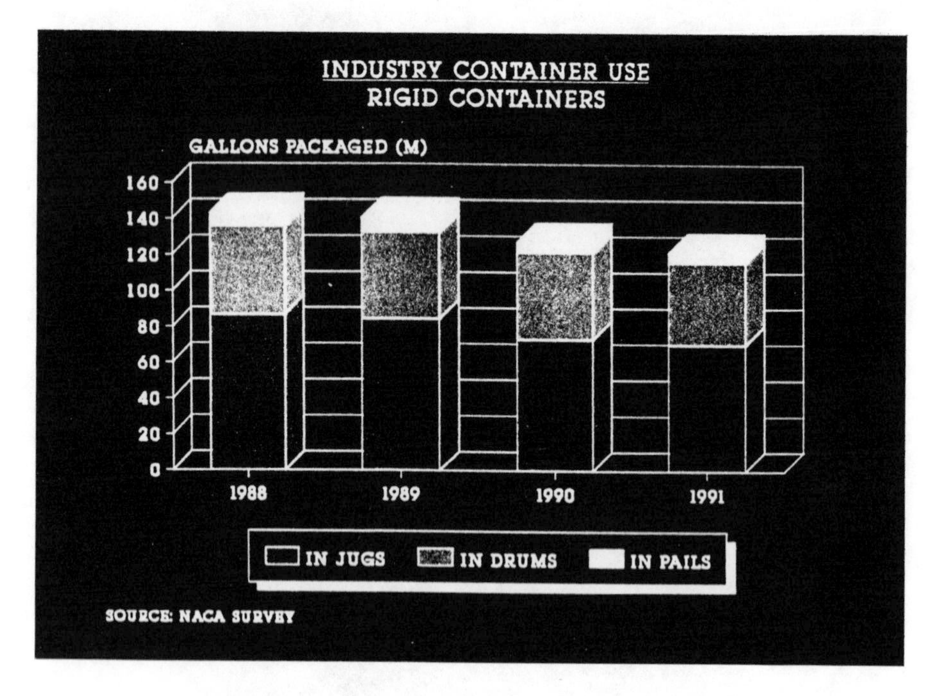

Figure 15. Rigid Single Trip Container Usage

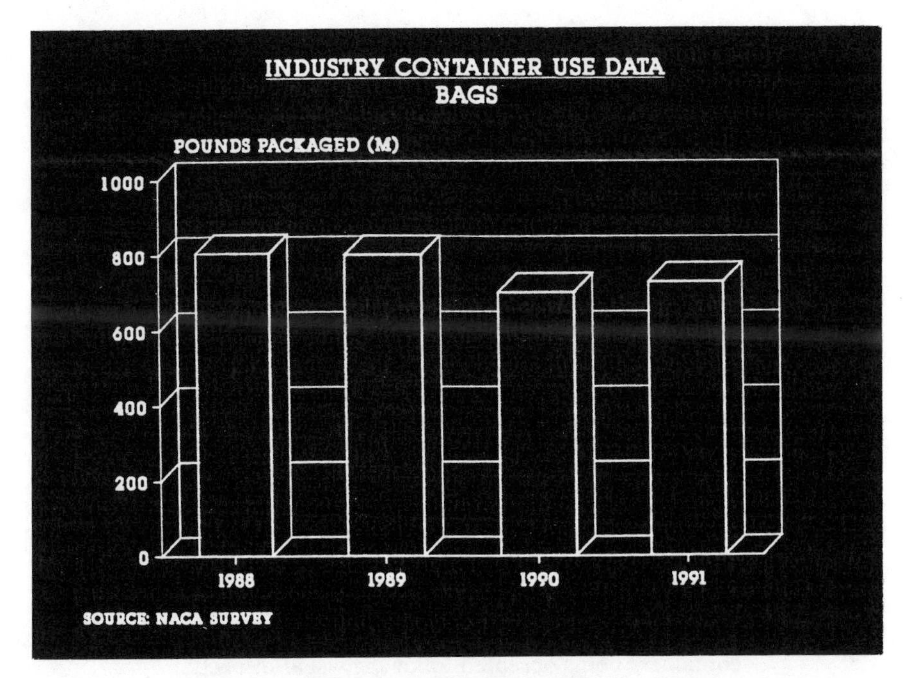

Figure 16. Flexible Single Trip Container Usage

While the industry's use of plastic and multi-wall bags for dry products has also declined, the decline is less dramatic than that for liquid containers (Figure 16). As was mentioned above the development of reusable containers for dry products has lagged behind that for liquids. It is apparent, however, that much attention will now be focused on this type of delivery system and progress in reducing our use of single trip bags is anticipated.

The major ag chemical producers have each set aggressive internal goals to reduce their use of one-way containers sharply by the mid-90's. NACA will continue to collect container usage data annually to measure the industry's progress in container source reduction.

To recap, the ag chemical industry is strongly committed to container source reduction. In meeting this challenge, it is pursuing several parallel approaches. These include the lightweighting packaging materials, the development of products that require less packaging, the exploration of new, potentially advantageous packaging technologies (like water soluble bags) and the introduction of reusable/refillable containers. During the 80's, the industry's use of one way containers has declined significantly, primarily because of the introduction of refillable liquid containers. Further reductions are anticipated as the bulk delivery of ag chemicals grows outside the Mid-West and as new types of refillable containers are introduced for both liquid and dry products. The industry has taken the initiative in developing design and performance guidelines to ensure that these new types of containers don't create unintended problems. These are accomplishments about which the agricultural chemicals industry can be justifiably proud.

RECEIVED May 22, 1992

Chapter 5

Pesticide Container Collection and Recycling in Minnesota

R. J. Hansen and L. P. Palmer

Agronomy Services Division, Minnesota Department of Agriculture, 90 West Plato Boulevard, Saint Paul, MN 55107

Minnesota's pesticide container collection and recycling pilot project collected data on 56,037 empty plastic and metal pesticide containers in twenty counties during 1991. High Density Polyethylene (HDPE) plastic pesticide containers comprised 94% of the total amount collected with 92% of the plastic containers being accepted following visual inspection. The project was designed to: collect, recycle and dispose of empty triple rinsed containers; evaluate current container management; and determine the cause and extent of the problems associated with pesticide containers. Seventy-one percent (71%) of the 662 participants in 1991 were farmers. The type and number of containers collected, survey information from participants, and residue analysis data will be presented. The number of containers, acceptance rates, quality, and number of participants increased in 1991.

Recycling of empty pesticide containers is a new and evolving technology. Collection systems, quality control mechanisms, and efficiency evaluation methods are still being developed. The Minnesota Department of Agriculture (MDA) has conducted various collection projects in an effort to determine the feasibility of empty pesticide container collection.

Pesticide Container Management in Minnesota

In Minnesota, options for proper disposal of empty pesticide containers have been reduced. The past acceptable practice of open burning is currently prohibited or strictly regulated *(1)*. Permitted solid waste landfills and municipal solid waste incinerators are refusing or limiting the quantity and

0097–6156/92/0510–0044$06.00/0

type of pesticide containers accepted. Recycling programs, relatively recent in rural and urban areas, are not monitoring the quality or quantity of pesticide containers entering their facilities.

Farmers, applicators and dealers surveyed. The 1987 Minnesota Legislature authorized the MDA to survey farmers, applicators and dealers about their attitudes and management practices regarding empty pesticide containers. Farmers surveyed were randomly chosen from a MDA state-wide list of certified private applicators. The resulting report *(2)* indicated that in comparison with other environmental issues in Minnesota, 21.9% of the 575 farmers responding to the 1987 survey rated empty pesticide container disposal as the most important environmental issue and 64.3% rated it an important environmental issue. Of farmers responding, 65% were using open burning to dispose of their empty containers.

Evaluation of Container Dump Sites. Pesticide container dump sites are a problem throughout Minnesota *(3)*. These dump sites occur where containers are discarded and accumulate over time. Containers found in such dump sites are often not properly rinsed and may result in soil and/or ground or surface water contamination.

Development of Proper Rinsing Education Campaign. In 1988, the MDA formed the Minnesota Pesticide Container Advisory Committee (MPCAC) to bring together farm, industry, environmental groups, the Minnesota Extension Service and other state agencies to develop the **Rinse and Win!** educational campaign to promote the proper rinsing of containers. The MPCAC developed a logo, fact sheets, public service announcements, posters and stickers to provide information about proper rinsing. This information was complemented by demonstrations showing and explaining the proper rinsing of containers. The information was prepared in time for the 1989 growing seasons and demonstrations occurred throughout the year. The campaign was modified to include a recycling component in late 1989 and promotional efforts on proper rinsing increased during the spring of 1990.

Authorization of Collection and Recycling Pilot Project. The 1989 Groundwater Protection Act authorized the collection and recycling pilot project as a two year, state funded effort to collect containers and gather information *(4)*. Additional provisions of the statute include a requirement for retailers to accept empty pesticide containers after July 1, 1994 *(5)*.

To assist with the project, the MDA continued the MPCAC and increased its membership. The MDA provided the project with one full time professional staff and a budget of $150,000 over the biennium from the Pesticide Regulatory Account, a registration fee based account. Organizations and individuals represented on the MPCAC also provided "in-kind" services and volunteers to the collection project.

A strategy for collecting, inspecting and accepting empty triple rinsed pesticide containers was developed by the MPCAC at a meeting on May 31, 1990.

Although late in the planting season, the MPCAC determined it was important to conduct several collections in 1990 to gain experience for more extensive collections in 1991. The MDA would conduct the collections using different methods in different regions of the state.

1990 Materials and Methods

Types of Containers Collected. Containers eligible for the collection included rigid plastic HDPE containers ranging in size from 0 to 55 gallons. Metal containers would also be accepted in the same sizes. Types of containers not included in the collection project included: paper, cardboard, glass, plastic low density polyethylene (LDPE) bags and pressurized metal containers. Containers larger than 55 gallons were also not eligible. Only containers which had contained pesticide products: insecticides, herbicides and fungicides were eligible for collection. However, empty crop oil, adjuvant and surfactant containers were also accepted. No distinction was made between agricultural pesticides and those used for home and garden use.

Only containers which had been properly rinsed (triple rinsed or pressure rinsed), were accepted for recycling. Visual inspection of individual container was required. Plastic and metal containers were checked for visible, colored residue (solid or liquid) on the interior and/or exterior of the container. Participants in the project were encouraged to bring in clean, dry, empty containers.

Some containers which did not meet this criteria were rejected at the collection site. The reasons for rejection varied, but were explained to the participant. The individual was provided with options that would allow for safe and proper disposal of the container. This included proper rinsing instructions and written instructions to apply the rinsate according the label directions or to store for future disposal.

Collection Logistics. Collections were held at agricultural chemical retailer sites, county fairgrounds, former Department of Transportation (DOT) facilities, and solid waste centers. Collection sites were required to provide shelter, electricity, rest rooms and concrete or paved surfaces. The sites were accessible to highways and identifiable by local residents.

Collection sites were distributed in varying regions within Minnesota; Houston county in southeastern Minnesota is a hilly region with karst geology and agriculture characterized by corn, legumes, beef and dairy; Isanti county in east central Minnesota is an urban/rural transition county with coarse textured soils characterized by truck farms, corn and assorted livestock; Stevens, Pope and Swift counties are in west central Minnesota with rolling glacial till and agriculture characterized by corn, soybean, small grain and beef and hog production; Polk, Pennington and Red Lake counties are located in the Red River Valley of northwestern Minnesota with lacustrine, alluvial, and glacial feature and agriculture dominated by small grains and sugar beets.

Process of Collecting of Pesticide Containers. Due to safety concerns, containers were inspected only by trained MDA employees. Inspectors were provided with nitrile gloves, Tyvek aprons or lab coats, goggles and required to wear steel toed boots. Volunteers who handled containers were also provided with personal protective clothing. Additional volunteers who collected data on the number, type, size and rejection rate of the containers did not come in contact with the containers and thus did not wear personal protective clothing. Incident response, police and fire department telephone numbers were posted prominently at the sites.

Plastic containers accepted following visual inspection had caps and labels removed. Caps were often polypropylene or poly-vinyl chloride (PVC) and labels were either LDPE or a combination of plastic, adhesive and paper. Aluminum foil safety seals were also removed if possible. The remaining container was granulated using a portable granulator which cut the plastic into 3/8 inch flakes. Granulating of containers was done outside under dry conditions or in a well-ventilated area. Granulated plastic was then deposited in a LDPE-lined cardboard box for transportation.

1990 Collection Results

Table I. Containers Collected and Accepted by Site

County and Dates	Containers Brought In	Containers Accepted
Isanti August 13 - 18	983 plastic 33 metal	782 plastic 22 metal
Pope September 12-14	1,061 plastic 131 metal	907 plastic 121 metal
Stevens September 12-14	617 plastic 166 metal	485 plastic 102 metal
Swift September 12-14	2,987 plastic 268 metal	2,618 plastic 162 metal
Houston May - September	4,096 plastic	3,102 plastic
TOTAL*	9,744 plastic 631 metal	8,676 plastic 429 metal

* Additional containers were collected in Red Lake, Polk and Pennington counties, but resulted in incomplete data, therefore, containers collected in their counties are not included in 1990 totals.

The granulated plastic was processed by Envirecycle Company Inc., a plastic recycling firm in Kansas City, Missouri. The processed plastic was used by E. I. Du Pont De Nemours & Company (Inc.) in the production of pesticide containers.

Metal containers accepted following visual inspection were crushed on site using a modified log-splitter for the five gallon or small cans and a large hydraulic can crushers for 30 and 55 gallon barrels. Metal containers were transported to a recycling firm as scrap metal.

The collection which was held in Houston county was locally conceived and operated. The MDA provided inspection of the containers at the end of the growing season. The collection was operated by county solid waste center and the plastic was shredded, then granulated using their own equipment. The Houston County collection was operated during the growing season (May -September) as compared to the one to three day collections operated by the MDA.

Participant Surveys. Participants were surveyed for: attitudes about container disposal and recycling, demographic information regarding size of farm and distance traveled, and pesticide container management practices. The surveys were anonymous and resulted in multiple answers for a number of questions. Survey results were summarized as percentage of responses.

Residue data. Residue data was also collected. This included rinsate samples collected from randomly selected, visually inspected and accepted empty containers. Rinsate samples were collected by inserting 0.5 L of deionized water from a 0.5 L amber bottle, capped with a Teflon lined cap, into the pesticide container. The pesticide containers were agitated for 30 seconds to rinse the inside of the container including the hollow handle. After rinsing, the rinsate was deposited into the amber sample bottle, capped and transported to the MDA Laboratory Services Division for analysis.

Composite plastic samples were also collected. A 0.5 L amber, Teflon capped bottle was filled with the mixed, granulated plastic. Methodology used for analysis was primarily the MDA Division of Laboratory Services s-triazines and non-acid pesticides screen, unless otherwise instructed. Pesticide analysis was performed by gas chromatograph (GC); laboratory quality control included confirmatory analysis.

1990 Observations. Eighty-nine percent (89%) of inspected plastic containers were accepted for recycling. All containers brought to the MDA operated collection sites were counted. This and other data, such as container type, size, and product information was recorded during the collection process. For example if a container was brought in, inspected and rejected, then properly rinsed, returned and accepted, it would be counted as two containers, one accepted and one rejected.

Sixty-eight percent (68%) of metal containers inspected were accepted for recycling. Metal containers were more difficult to inspect because of the age, rust and deterioration of the containers. Metal containers often required the use of a flashlight to determine if any residue remained in the barrel or can because it was difficult to distinguish between residue and rust.

In all, eighty-eight percent (88%) of the containers brought into the collection were accepted for recycling. This was interpreted as an acceptable rate for the first year of the collection. However, the total number of containers collected was relatively low for the potential of containers to be brought in. Approximately 100 farmers, applicators or dealers participated in the project.

1990 Rinsate Analysis Results

Table II. Pesticide in Rinsate of Visually Inspected Containers

Container Brand and Type	Pesticide, Concentration mg / L	% Removal
Lasso, plastic 2.5 gallon	Alachlor, 31	99.9993
Dual, plastic 2.5 gallon	Metolachlor, 9.2	99.9999
Bicep, plastic 2.5 gallon	Metolachlor, 0.70 Atrazine, 0.07	99.9999
Lasso, metal 5 gallon	Alachlor, 2.7	99.9998
Sencor, plastic 1 gallon	Metribuzin, 40	99.9978
Malathion, metal 5 gallon *	Chlorothalinol, 0.11 Malathion, 0.01 Methyl Parathion, 0.06	100.00

* Confirmation: GLC 30m DB-1, 17 TSD / ECD, 30m DB - 5 GC / MSJ

The inspection process was very labor intensive. MDA had one or two staff on site to inspect all containers resulting in a "bottleneck" in inspection and increased time for processing the containers. Collections occurred late in the growing season after many containers may have been disposed of by other means. When those who participated in the collection were asked "How do you normally dispose of your pesticide containers?", 27% replied "burning on the farm"; 19% "take to the local landfill"; 17% "store on the farm"; 11% "return to dealer"; 11% "use for something else".

1991 Collection Strategy

Increased Educational Efforts. The MPCAC met on December 4, 1990 to plan for the 1991 collections. Goals for 1991 were to: continue the project in the 1990 regions, increase participation while lowering the rejection rate, expand into regions of the state not included in 1990, examine more local control of the collections, and intensify educational efforts about proper rinsing and container recycling.

Local planning committees were established in each region to provide ideas and suggestions on meeting the local needs for the collection. Contact persons, most commonly county extension agents (12 out of the 20 local coordinators), were identified in each local planning group to coordinate information and education efforts.

Recycling demonstrations, using the container granulator, were scheduled at farm shows during the winter. Display booths and information about the collections were provided by the MDA to local coordinators. Posters, brochures and other handouts were designed to provide information about each collection. Newspaper articles, direct mail, extension bulletins, dealer handouts, radio and television were also used to get information to potential participants as to when and where the collection would be held and how to participate.

An Operations Manual was developed to describe local and MDA responsibilities for container collections *(6)*.

1991 Materials and Methods. Inspection methods remained the same. However, the procedure was defined and described in an Inspection Manual *(7)* developed to train local personnel. MDA staff would no longer need to inspect every container as local personnel would be trained to provide this role. Volunteer responsibilities were also further defined.

The plastic containers were granulated using a <u>Cumberland</u> granulator which was newer and had greater capacity to process containers than the one used in 1990. The granulator was mounted on a flatbed trailer and transported from site to site for use at individual collections. The granulator required 220 volt, 3-phase electricity to operate. The granulated plastic was blown through an enclosed system and deposited into a plastic "super-sack" which held approximately 1,000 pounds each and served as a storage unit for transportation to final destination.

Container sampling procedures and analytical methods remained the same. Rinsate samples would comprise most of the samples collected in 1991. However, some composite plastic samples would also be collected to determine the effectiveness of visual inspection. Methodology and confirmation of analysis remained the same as in 1990.

Collection Logistics. Twenty counties were chosen for container recycling in 1991, Figure 1. The self-operated style of collection used in Houston county was also adapted for neighboring Fillmore and Winona counties. In Fillmore county the containers were collected during the month of May at the county recycling center. Recycling center staff were trained by the MDA to inspect containers. In Winona county, containers were returned to the dealerships and consolidated by the county solid waste officer in August at a privately operated recycling center. Houston county continued its program similar to 1990, but with increased educational efforts.

A collection for Martin county was scheduled in late May. Martin county is located in south central Minnesota and has an agriculture dominated by corn, soybeans and hogs. This collection was followed by a joint collection between Martin and Jackson counties in July. This collection was also a combination of the MDA's waste pesticide program and container collection pilot project. Both collections were held at the county fairgrounds.

The 1990 collection in Isanti county was followed in 1991 by single day collections in Isanti and four surrounding counties: Pine, Chisago, Kanabec and Mille Lacs. The collections were operated at county DOT or solid waste facilities. These five counties have solid waste management plans operated by the East Central Solid Waste Commission.

The agricultural chemical dealerships of McLeod county initiated and operated a two day collection in July at the county DOT facility. The MDA provided training for volunteer inspectors. McLeod county is located in central Minnesota and is characterized by corn, soybean, dairy and hogs.

Collections in Stevens, Pope and Swift counties were operated at the same time, date and location as in 1990. Educational efforts were intensified as planning for the collection occurred prior to the spring application season.

The collection for Polk, Pennington and Red Lake counties in 1990 was followed by a greatly expanded series of collections in 1991. Individual collections were operated in Norman, Polk, Pennington, Red Lake, Marshall and Kittson counties. In Norman county three collections were held at three sites (one day at each site) at county DOT facilities. In Polk county, the MDA trained recycling center staff to inspect containers and the collections occurred from early May until the middle of July at the county solid waste transfer station.

In Pennington, Red Lake and Marshall counties, collections occurred at two sites in each county. The collection sites were either agricultural chemical dealerships or county DOT facilities.

The Kittson county collection was held at the county fairgrounds for one week in late June. Container inspection responsibility was shared between both MDA staff and trained local inspectors.

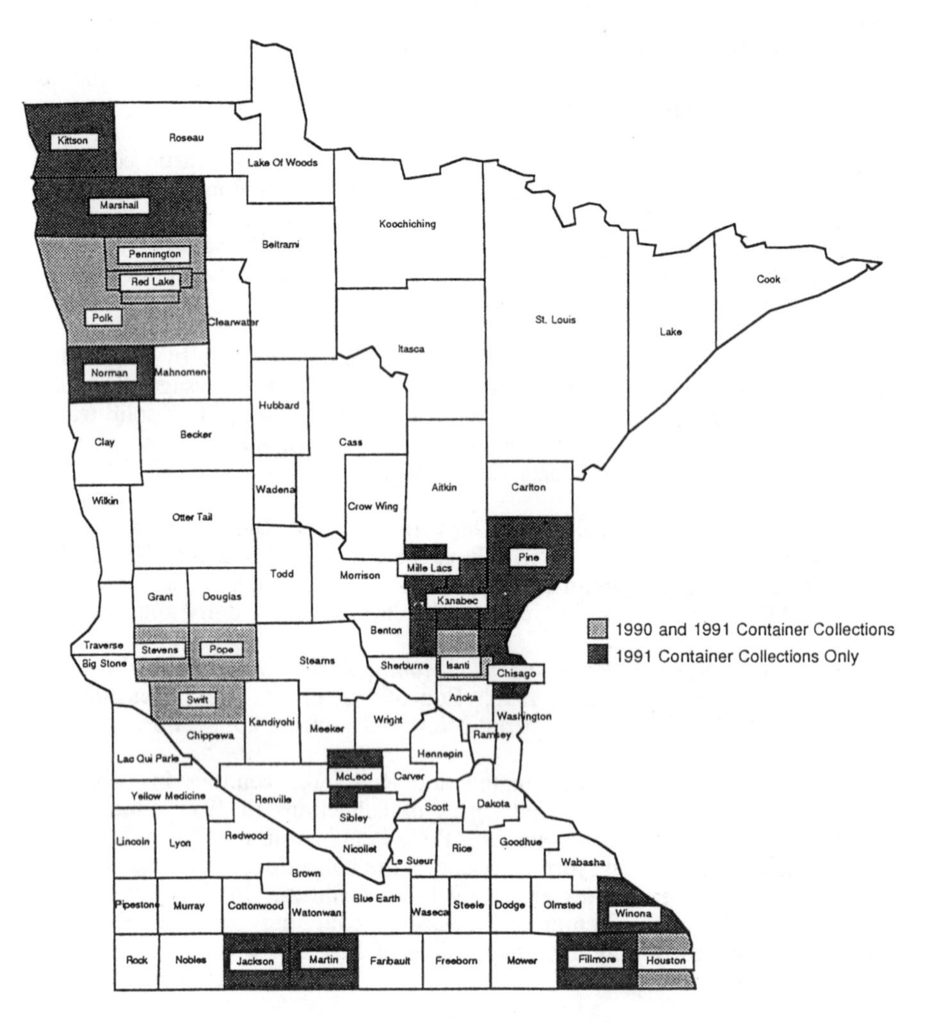

Figure 1. Minnesota Counties with Pesticide Container Collections

1991 Collection Results

Table III. Containers Collected and Accepted by Site

County and Dates	Containers Brought In	Containers Accepted
Pine, Isanti, Chisago, Kanabec, Mille Lacs July 15-19	4,500 plastic 81 metal	4,016 plastic 80 metal
Martin, Martin/Jackson May 29-31, July 24-25	4,088 plastic 649 metal	3,771 plastic 552 metal
Winona June - August	1,010 plastic	883 plastic
McLeod July 16-17	3,098 plastic 548 metal	2,776 plastic 541 metal
Polk May - July	10,846 plastic 233 metal	10,739 plastic 230 metal
Norman July 8-10	9,339 plastic 832 metal	8,237 plastic 697 metal
Pennington June 10, 15	1,065 plastic 92 metal	921 plastic 80 metal
Red Lake June 12, 14	1,976 plastic 254 metal	1,723 plastic 104 metal
Kittson June 24-28	3,576 plastic 367 metal	3,304 plastic 230 metal
Marshall June 17, 19	4,681 plastic 281 metal	4,157 plastic 255 metal
Pope, Stevens, Swift September 11-13	8,255 plastic 266 metal	7,914 plastic 255 metal
TOTAL*	52,434 plastic 3,603 metal	48,441 plastic 3,024 metal

* Additional containers were collected in Fillmore and Houston counties, but resulted in incomplete data, therefore, containers collected in their counties are not included in 1991 totals.

Overall, planning for the 1991 collections occurred earlier than 1990, with planning meetings at the local level taking place in January, February and March. Early planning allowed time for the local coordinating groups to get information about the collection to local growers and applicators prior to spring planting and application season.

Site selection within the counties was also completed earlier in the year allowing potential participants to know when and where the collections would occur thus encourage saving of containers.

1991 Observations. Ninety-two percent (92%) of the total inspected plastic containers were accepted for recycling. Containers were counted in the same manner as 1990. Eighty-four percent (84%) of inspected metal containers were accepted for recycling. Metal container inspection had the same difficulties as in 1990; rust and deterioration. At the collections, MDA personnel provided most of the metal container inspections.

Excluding Fillmore and Houston counties, ninety-two percent (92%) of all the containers brought into the collections were accepted for recycling. Counties where collections were held two consecutive years had an increase in number of containers received and level of participation. Statewide the number of participants rose to 662 in 1991 compared with approximately 100 in 1990.

Electrical requirements for the granulator were difficult to fulfill in the rural areas where the collections took place. Most of the sites did not have 220 volt, 3 phase electricity to operate the machine. This resulted in additional costs to prepare for and conduct the collection. Also, hand "feeding" the granulator was labor intensive and cumbersome. A mechanical feeding system and a self contained power source would improve the granulating process.

Rinsate and plastic residue analysis results for 1991 have not been completed.

1991 Survey Responses. Of the participants who completed surveys at the collection, responses on normal disposal of containers were similar to results from 1990 participants. Burning on the farm was the answer with the highest percentage of responses, Figure 2. Participants could give multiple answers to the question. The questionnaire was a two-sided, multiple-choice survey.

When asked, "How should pesticide containers be managed?", burning, burying and land filling were not endorsed by those participating in the collections, Figure 3. Some form of return program was strongly supported. Collection/recycling programs; returned to dealer; returned to the manufacturer; reused/refilled or deposit/return programs are examples of choices available or suggested programs.

The desire to recycle and convenience were important reasons given for participating in the collection project, Figure 4.

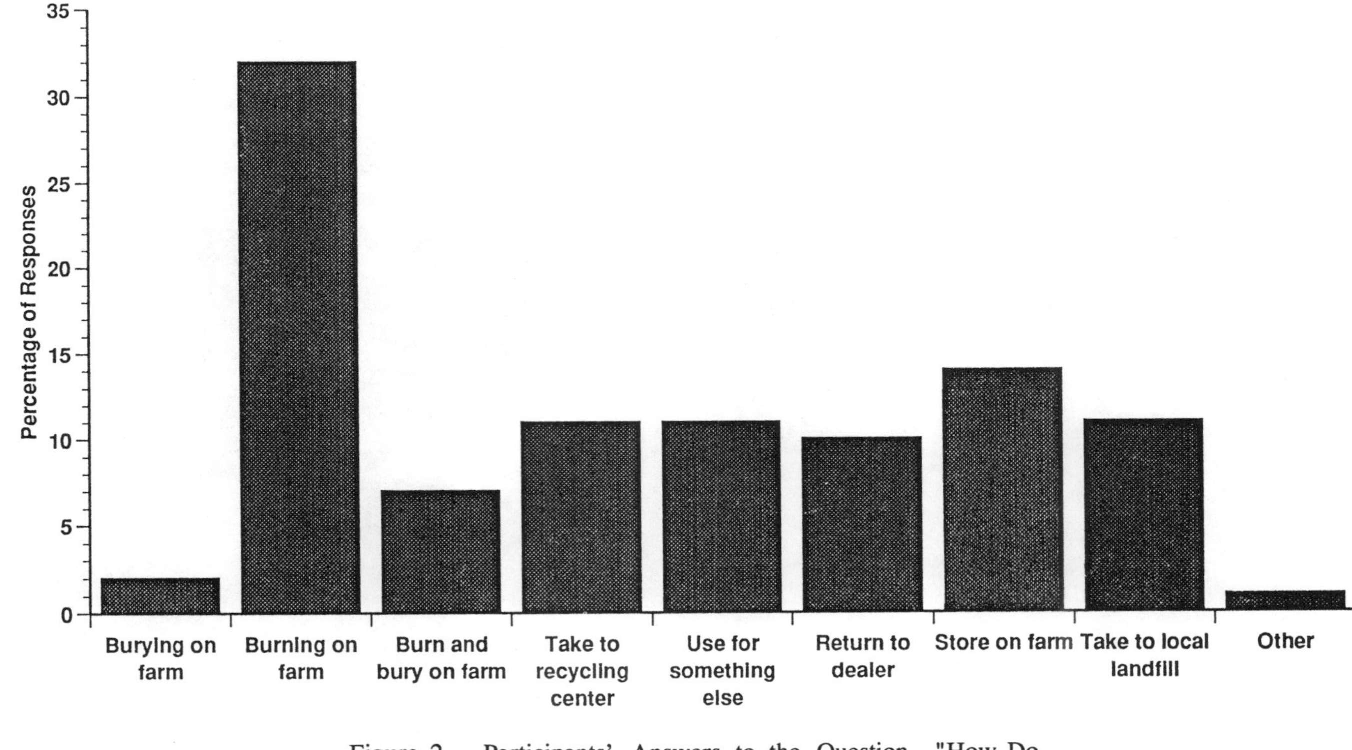

Figure 2. Participants' Answers to the Question, "How Do You Normally Dispose of Your Pesticide Container?" The statewide composite data in Figures 2 - 7 is from 1991 Minnesota Department of Agriculture container collections excluding Fillmore, Houston and Winona counties

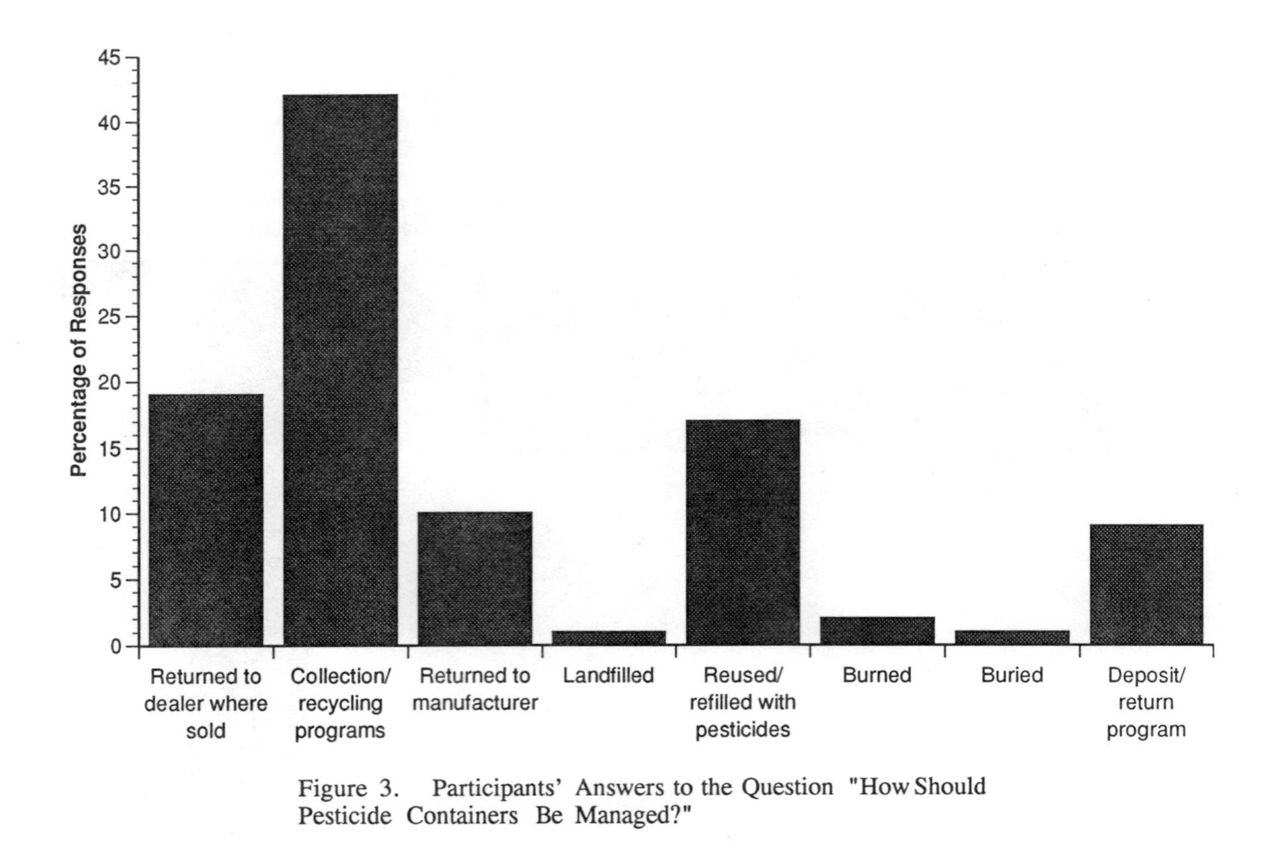

Figure 3. Participants' Answers to the Question "How Should Pesticide Containers Be Managed?"

Figure 4. Participants' Answers to the Question, "Why Did You Participate in the Collection Project?"

In 1991, educational efforts were increased to provide information about how to participate in the collection project. A variety of methods were used to provide information. Participants acknowledged this multiple media effort by listing many of the methods, Figure 5. Newspaper articles and other printed media were effective in reaching those who participated in the project.

Most of the participants in the collection project traveled greater than five miles, Figure 6. Also, most of the farmers participating came from large farms, Figure 7.

Conclusions

The pesticide container collection and recycling pilot project developed a variety of feasible collection models for groups, business or local government to use in collecting containers. Visual inspection of pesticide containers was an effective way of determining whether the container had been properly triple rinsed and was free from visible residue. Participants supported some form of collection or return program to manage empty pesticide containers.

Local coordination and cooperation are essential in providing information about the project and achieving participation. Multiple education efforts by various groups, organizations and individuals enhance the likelihood of success of the collection effort. Our experience indicates that a strong local coordinator such as a county extension agent, who is knowledgeable about their area and has the respect of potential participants is an essential component to a successful collection.

Limited time (one to three day) collections were as effective in collecting similar quantities of containers as on-going collections. Limited time collections also had intangible benefits of having various groups cooperate in an environmental/agricultural project. However, limited time collections may be more inconvenient for participants and require greater MDA and local resources than ongoing collections. Potential problems with storage, exposure, and handling may be less with short term collections. On-going collections may be more convenient for participants, but increase potential problems with staffing, storage, exposure, and handling.

Plastic containers were the most common type of container collected. Metal containers were more difficult to inspect and had a higher rejection rate than plastic containers. The inspection process, though effective, was time intensive, requiring commitment of both MDA and local resources. Data collection, removal of labels and caps, and handling of the containers occupied most of the remaining time. The inspection process resulted in very "clean" containers being accepted for recycling. Rinsate analysis showed greater than 99.99% removal of pesticide from sampled container. Defined procedures for inspection and consistency were important components in working with participants in this voluntary program. In 1991, more containers were brought into the collections, by more participants, with a lower rejection rate.

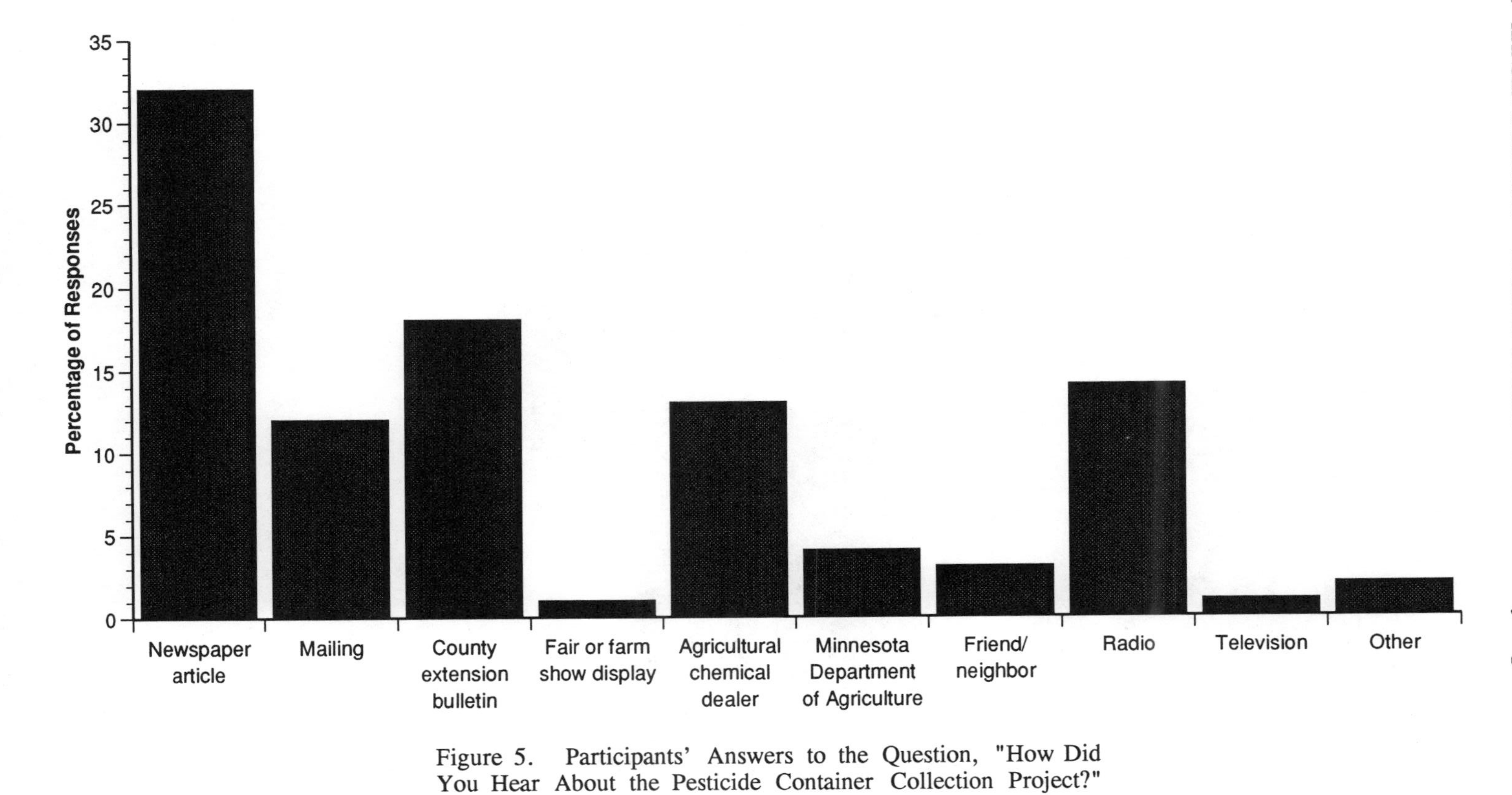

Figure 5. Participants' Answers to the Question, "How Did You Hear About the Pesticide Container Collection Project?"

Figure 6. Participants' Answers to the Question, "How Far Did You Travel Today?"

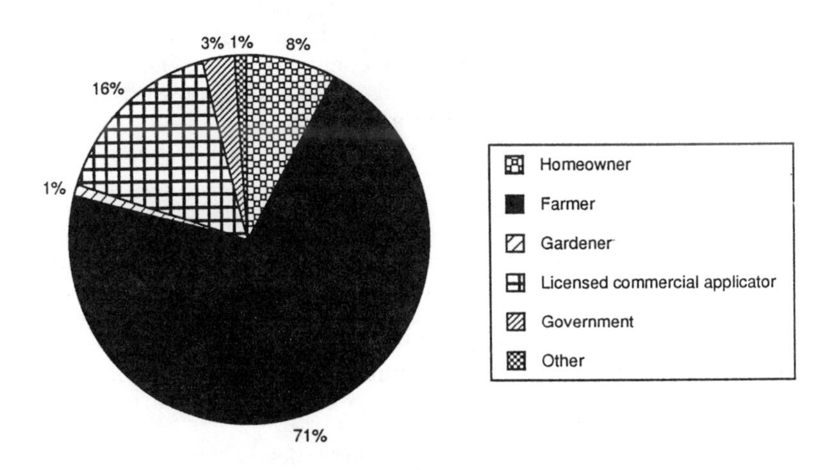

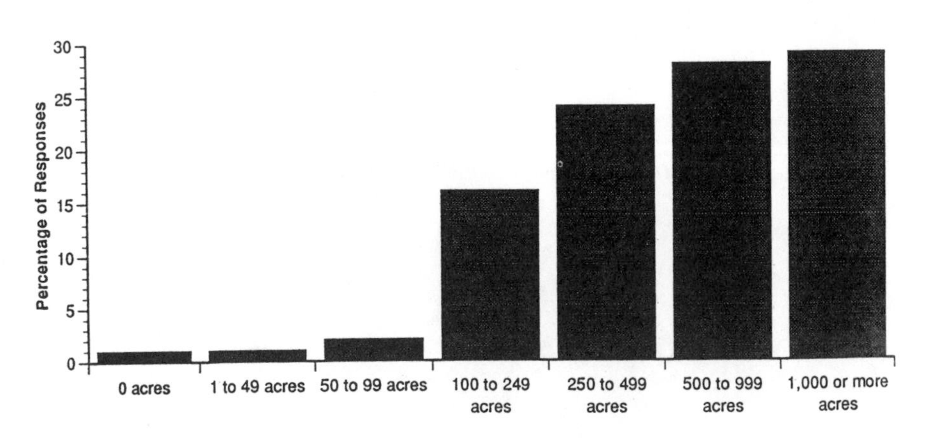

Figure 7. Participants' Answers to the Question, "If the Operation is Farming, Please Indicate Your Farm's Acreage?"

Pesticide container collection projects were very popular with those who participated. Pesticide users (farmers, applicators or dealers) who participated in the project desire methods of pesticide container management different from those currently used. Recycling, returning, and reusing were popular suggestions for programs in the future.

Based on observations made and data collected during the 1990 and 1991 pesticide container collection and recycling pilot project, the MPCAC at a meeting on October 8, 1991, unanimously recommended the MDA to continue and expand pesticide container collections in 1992 and a evaluate possible statewide expansion of the project into a program.

LITERATURE CITED

(1) Laws of Minnesota for 1989; Senate File No. 281; Chapter 131 Section 2 (17.135) Farm Disposal of Solid Waste.
(2) Minnesota Department of Agriculture; Minnesota Empty Pesticide Container Disposal Report; St. Paul, Minnesota, March, 1988.
(3) Minnesota Department of Agriculture, Greg Buzicky; Evaluation of Improperly Disposed of Pesticide Containers on Minnesota Farms; St. Paul, Minnesota, October, 1990.
(4) Laws of Minnesota for 1989; Chapter 326, Article 5, Section 52; Pesticide Container Collection and Recycling Pilot Project.
(5) Laws of Minnesota for 1989; Chapter 326 Article 5, Section 29; Sale of Pesticides in Returnable containers and management of unused portions.
(6) Operations Manual; Minnesota Department of Agriculture, April, 1991
(7) Inspection Manual; Minnesota Department of Agriculture; April, 1991

RECEIVED August 3, 1992

Chapter 6

Laboratory Evaluation of Products of Incomplete Combustion Formed from Burning of Agricultural Product Bags

B. Adebona, A. Shafagati, E. J. Martin, and R. C. Chawla[1]

School of Engineering, Howard University, Washington, DC 20059

Unused and used empty, aluminum-lined, multiwall bags, utilized as containers for pesticides were burned in an infrared incinerator in the induced air mode, at temperatures ranging from 300 to 1000 °C. Combustion parameters for test conditions were selected to simulate "open burning," a popular disposal option for used pesticide bags. Emissions were analyzed for organic products of incomplete combustion (PICs) using appropriate sampling equipment and a gas chromatograph / mass spectrometer (GC/MS). Preliminary analysis shows the presence of cyclic compounds in the emitted gas streams from burning both used and unused bags. Number and concentration of PICs decreased as the combustion temperature increased. Results illustrating the relationship between PIC formation and combustion temperatures are presented and discussed.

With the passage and implementation of the 1988 Amendments to the Federal Insecticide, Fungicide and Rodenticide Act (FIFRA-88), the U.S. Environmental Protection Agency (EPA) is proposing to revise the regulations related to storage, disposal, transportation and recall of pesticides and pesticide containers, and incorporate existing authority into 40 CFR Part 165. Associated data requirements will be incorporated in 40 CFR Part 158. Of the four elements of planned attention by EPA: storage, disposal, transportation and recall, **disposal** is of interest in this paper.

[1]Corresponding author

0097–6156/92/0510–0063$06.00/0

The primary options potentially useful for disposal of bags are recycle (probably not practical for paper bags), landfill (which implies costs and risks for transportation and long term storage), and open burning at the point of use. Open burning is the most used option because it is least costly to the user and is available on-site.

Many pesticide users are involved with disposal of the 1/4 billion pesticide containers generated each year (1). About 30% of the total number of containers is estimated to be 50-lb. bags (2). This constitutes use of about 75 million 50-lb. pesticide bags per year.

Brief History of Multiwall Shipping Bags

Over 100 years ago, shipping sacks first used to transport products such as flour and feed to the market place were made of burlap and/or cotton. Prior to this time, the popular method of shipping bulk materials was in barrels or wooden crates. When cotton became scarce during the Civil War, manila rope paper made from 100% manila fiber was introduced as a substitute for burlap and cotton. Bags were closed by gathering at the top and tying with cord. Early in the 20th Century a manila rope paper bag was invented; it was pasted at both ends with a "valve" opening. The bag was originally intended to package salt but was soon used for cement and limestone.

When manila rope became scarce after World War I, kraft paper was mixed with manila fiber, but the combination resulted in a very stiff composition. The idea of using two plies, each of a lighter weight, was introduced to give a more acceptable bag construction. The multiwall bag became a reality. In 1924, 3, 4, and 5 wall bags were introduced, but made entirely of 100% kraft paper. The problem of closure was solved by sewing the plies together. Today the multiwall bag is very sophisticated and contains many other materials besides paper to provide properties of strength, moisture resistance, oil penetration resistance and odor prevention (Figure 1). Kraft paper may be laminated with low and high density polyethylene, aluminum foil, and glassine (3). Bags may be coated with various substances like inks and dyes that are used to provide a wide range of data and information to the user. Thus today's shipping bag may be a complex combination of chemicals in addition to the contained material(s), adding to the complication of assessing the environmental and health consequences of bag disposal.

Why be Concerned with Bags?

Product retention studies by container manufacturers have indicated that some product is retained in the bag after emptying. The average quantity of 5% Diazinon formulation, for example, retained in the bag in one study following

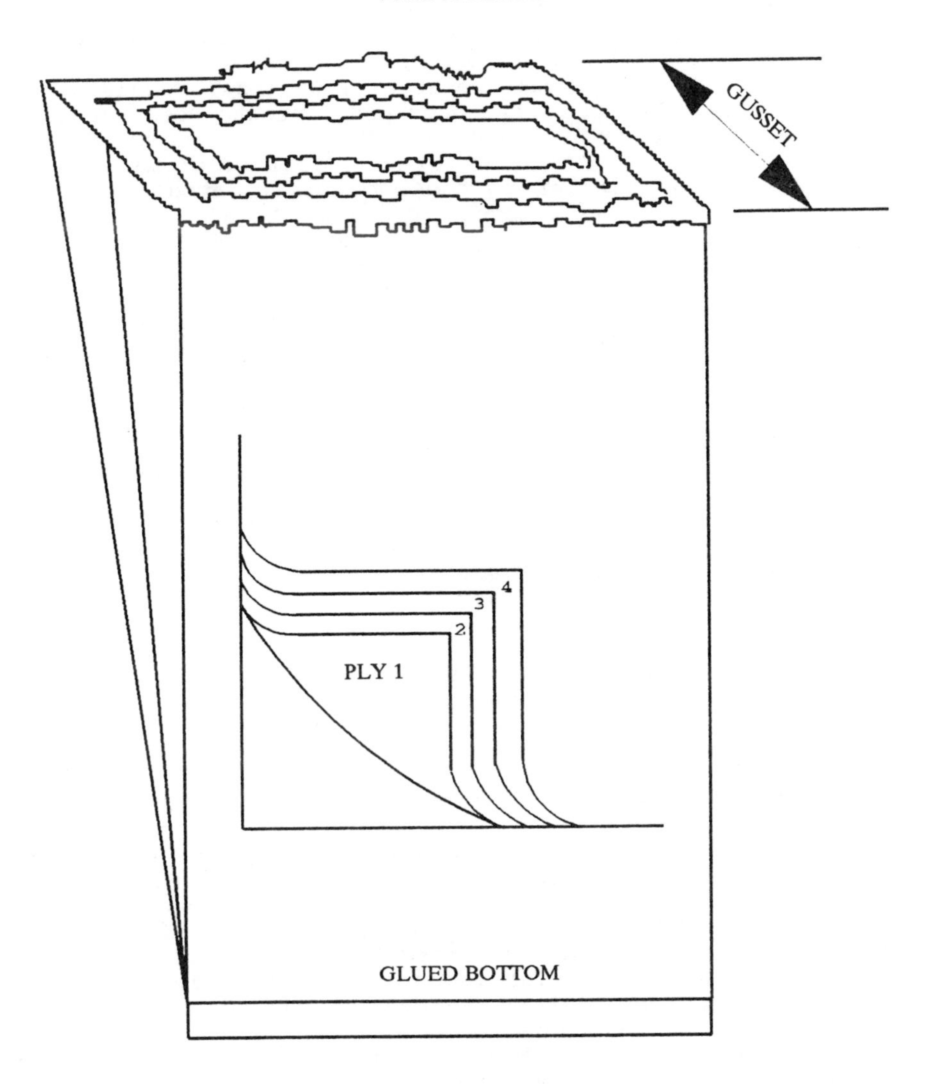

PLY 1 = ADHESIVE + METALLIZED FILM
PLY 2 = ADHESIVE + PAPER
PLY 3 = ADHESIVE + PAPER
PLY 4 = PAPER + INK

Figure 1 Multiwall Bag

standardized opening, emptying and shaking tests, was about 0.12 grams or 2.64 x 10^{-4} lb. (Letter from E. Tytke to B. Omilinsky of Formulogics on Stone Container Corporation Product Retention Evaluation, August 7, 1990). In a 50 lb. bag this represents about 33.6 x 10^{-6} lb of Diazinon. Using this value, the amount of Diazinon remaining in bags on a nationwide basis would be about 2600 lbs.(roughly one ton) annually. If all of the bags and contents were burned at 99% destruction efficiency, about 25 lbs./year would be discharged to the atmosphere. Other estimates of material retained in bags similar to the one investigated in this study range as high as 2 grams or 4.4 x 10^{-3} lb. (3).

In addition to the bag contents - in this case a pesticide - the bag and its construction components may be sources of PICs. Table I shows the possible sources of PICs among the bag components. The bag used for this study contains only the components shown with an asterisk (*).

A risk analysis for incineration of used pesticide containers has not been performed as yet. It would provide an indication of the potential impact of the residual pesticide (in this case Thimet) material left in containers. The risk analysis would proceed in several estimation steps:

- residual pesticide in container
- number of containers used
- number of containers burned
- types and quantities of emissions based on this and other combustion studies
- potentially exposed population
- dispersion to the exposed population

The Howard Program

The Howard University Combustion Research Laboratory (HUCORL) is dedicated to the analysis and discovery of the requirements for effective combustion of wastes and the understanding of the combustion process. HUCORL was established to develop and refine combustion technology and its applications to all industrial and commercial processes that require burning as an integral practice. Because hazardous waste destruction is an important national need at this time, the primary thrust of the work of the Laboratory is directed at this area.

Incineration is a destruction technology that has been developed for the permanent disposal of hazardous wastes. On the other hand, open burning of pesticide containers represents uncontrolled combustion without the advantages of air pollution control. One of the combustors available at HUCORL, the infrared unit can be operated in a manner to simulate open burning, i.e., the temperature profiler can be set to range from room temperature to a preset maximum. In the case of low preset temperatures, e.g., 300 ^{o}C, the temperature in the furnace will sometimes exceed the preset

TABLE I

POTENTIAL SOURCES OF PICs

LAMINATES

- * PAPER
  - HDPE - high density polyethylene
  - LDPE - low density polyethylene
- * METALLIZED FILM
  - * MYLAR
  - POLYPROPYLENE
  - NYLON

* ADHESIVES

* INKS

* DYES

GLASSINE

* ADDITIVES - for strength, durability, appearance, etc.

(classified chemicals, in general)

maximum when a flame is present. Different settings of the maximum temperature during testing can be established in order to simulate the exposure of various segments of a pile of unburned bags, as might occur in a field where disposal by burning is being practiced.

Samples of actual multiwall bags that had contained a granular formulation of phorate (Thimet 10G/15G; diethyl S-[(ethylthio) methyl] phosphoro-dithioate) were sent to Howard University by EPA after being emptied using normal field practices. The bags were sampled by cutting various portions which might represent different quantities of residual Thimet formulation. These samples were then burned in the infrared combustor under various operating conditions selected to cover the range of conditions expected during open burning conditions. Table II presents a range of commonly used pesticides along with Thimet.

Experimental Setup

The primary component of the infrared combustor is a small scale (100 gram capacity) electric powered insulated and controlled furnace shown in Figure 2. It is possible to record the weight change of a sample during and after combustion using a balance on the top of the unit. The furnace is also equipped with a temperature recorder. The infrared unit is related to other ancillary equipment in the laboratory as shown in Figure 3.

The temperature profiler controls the rate of temperature increase during a burn and the maximum temperature that may be preset. An afterburner with a separate temperature controller insures that any residuals which are created in the furnace are burned to a high degree of efficiency before hot gases exit the lab through the building stack (dedicated exclusively to HUCORL). Induced rather than forced draft provides that any leaks in the system are into the duct work and afterburner and finally through the stack rather than into the room. Sampling is performed from the duct work between the furnace and the afterburner; volatiles are sampled using the volatile organic sampling train (VOST), and semi- and non-volatiles are sampled using Modified Method 5 (MM5 or Method 23). MM5 is the method used as a "standard" technique for years to gather particulate and non-particulate, non-volatile compounds from air samples using a combination of collection in liquid and adsorption on solid medium. The VOST is similar but used for sampling volatile compounds. The two systems often exhibit some "crossover," i.e., both volatile and non-volatile compounds may be collected on each.

Simulating an Open Burn

The profiler brings the temperature in the furnace up to the preset value within a few minutes as shown in Figure 4,

TABLE II

PESTICIDE TYPES

TYPE	EXAMPLES
ORGANIC	
Phosphorus-containing	THIMET (Phorate) Parathion Malathion
Nitrogen-containing	Carbamates Alachlor Captan Diquat
Sulfur-containing	Chlorobenside EXD
Chlorophenoxy	CPA 2,4-D
Polyhalogen	DDT DDD
Polygen-Aromatic	Toxaphene
ORGANO-METALLICS	Carbon-metal bonded
INORGANICS	
Heavy Metal-High Tox	Arsenic, Cadmium
Heavy Metal-Mod. Tox	Copper
Fluoro Compounds	Sodium fluoride
Miscellaneous	Al, Zn, Cyanide

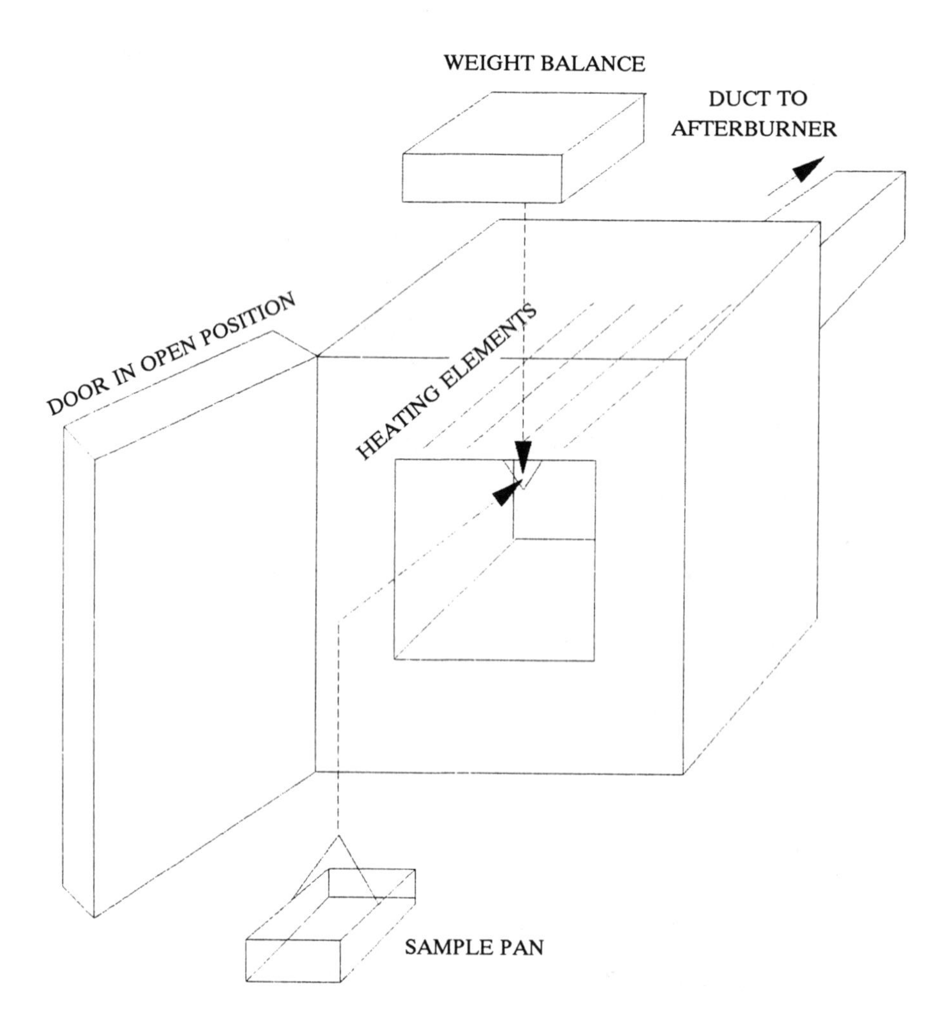

Figure 2 Infrared Furnace

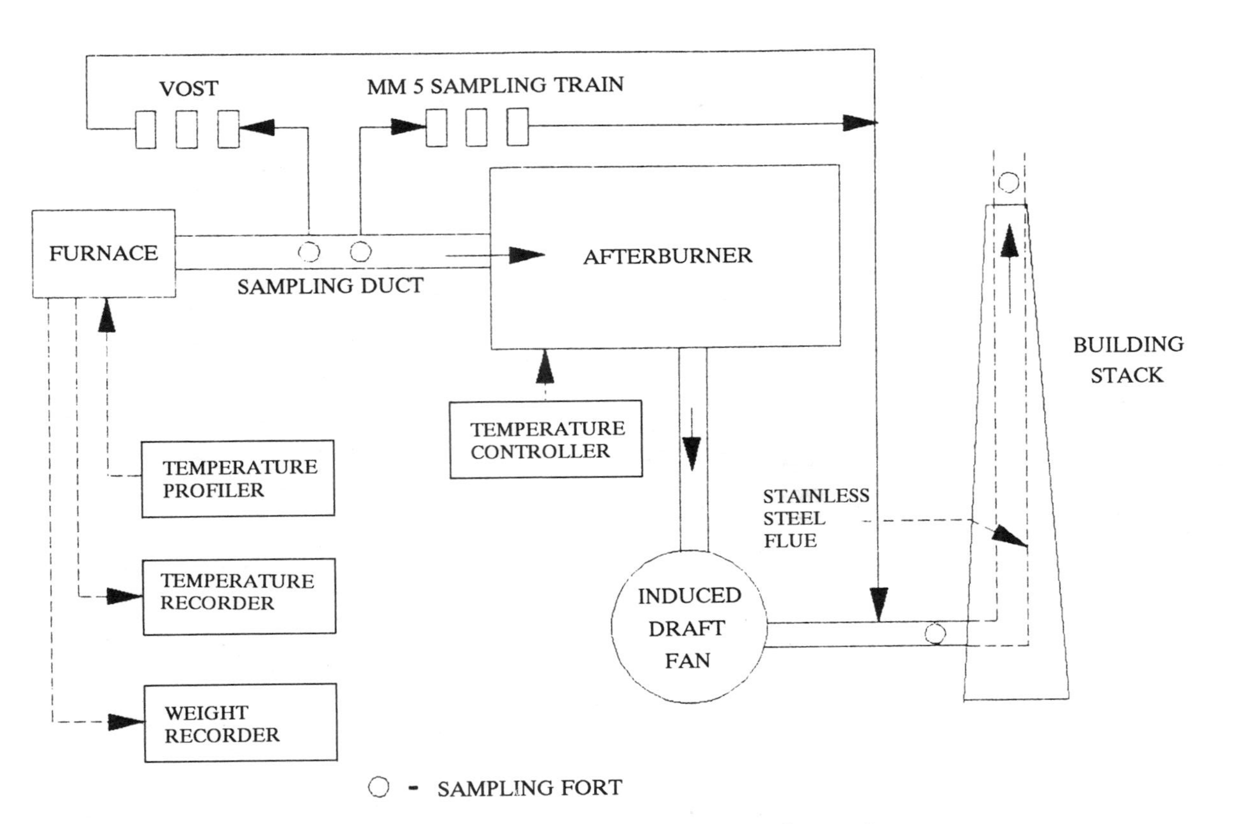

Figure 3 Combustion System Schematic

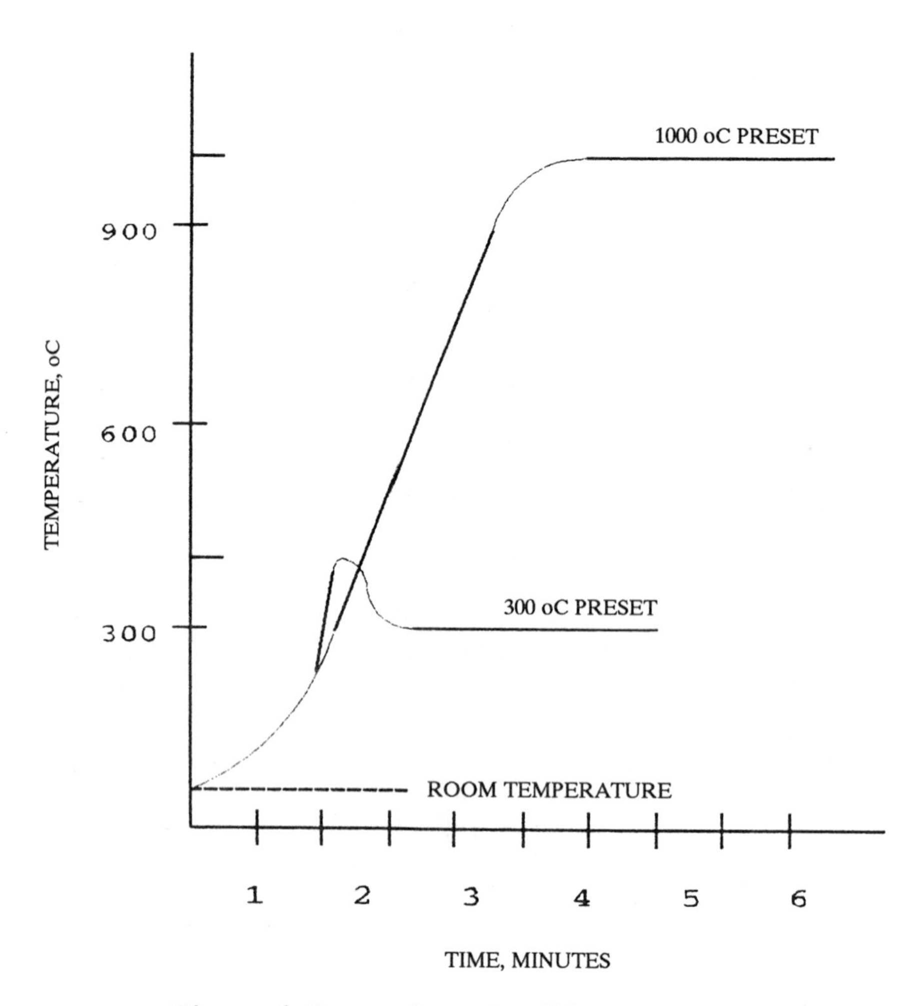

Figure 4 Temperature Profiler Behavior

regardless of the temperature setting. Referring to Figure 4, at lower temperature settings (300 °C in this case), the flame temperature as the sample burns may exceed the preset profiler value. The VOST and MM5 samplers may be operated for about 20 minutes in a given cycle; the sampling period therefore, extends beyond the completion of a sample burn during a test (typically 5 to 15 minutes). The collected sample contains a portion of the emissions from the entire sample burn. PICs produced at lower temperatures are thus a part of the sample as well as PICs from higher temperatures. Indeed, there are similarities among PICs at both high and low temperatures as discussed in the results.

The overall time period of a simulated burn was representative of an actual open burn. Within a few minutes of lighting bags with a match in the field, flames spread to engulf the entire pile and then recede. Most of the bag material is completely burned, some is charred (partially burned), and some is left unburned. Samples recovered after a simulated test burn showed similar characteristics.

Results and Discussion

On a preliminary basis (before confirmation of the identity of all compounds), the following total numbers of PICs were found by GC/MS analysis using the National Institute of Standards and Technology (NIST) library of reference spectra for comparison:

PICs from unused bags at 300 °C = 7
PICs from unused bags at 1000 °C = 8
PICs from used bags at 1000 °C = 14

More PICs were found from burning used bags than unused bags. About the same number of PICs (although some qualitative differences exist) were found from unused bags at both 300 °C and 1000 °C. This indicated as pointed out earlier, that all PICs produced during temperature profiling over the full range are gathered by the sampling systems. Almost all of the compounds - from used and unused bags - were polycyclic aromatic (multiple benzene ring) compounds in different configurations and with different substitutions, as shown in Table III. The presence of these compounds suggested origins other than the paper bag itself, or even the phorate (Thimet) molecule. Additional burns are concentrating on the inks and adhesives used to laminate the bag layers during construction.

On the other hand, the sources of PICs may be more difficult to determine. Table IV compares some of the PIC compounds from this study to two others. A complementary study done by Science Applications International Corporation (SAIC) for EPA used a full scale open burning simulation of Thimet bags in Florida (4). The HUCORL

TABLE III

SUMMARY OF PICs FROM BAG BURNS

COMPOUND	UNUSED 300 oC	UNUSED 1000 oC	USED 1000 oC
Napthalene	X	X	X
Azulene	X	X	X
1,3,5,7-cyclo-octatetraene	X	X	
Substituted Alkyl Benzenes	X	X	
2- dodecene	X		
2-methyl-2-propenylbenzene	X		
Indenes		X	
Cyclotetradecane		X	
Acenapthalene			X
Diethylbenzene			X
Decahydro-diethylnapthalenes			X
Anthracene			X
Phenanthrene			X
9H-fluorene			X

TABLE IV

Confirmed Compounds Recovered from Burns of Bags and Plastic Mulch Material

COMPOUNDS	TYPE	HUCORL	SAIC	LINAK
Naphthalene	polycyclic	X	X	X
Biphenyl	"	X		
Acenaphthalene	"		X	X
2-methylnaphthalene	"		X	
Flouranthene	"	X	X	
Phenanthrene	"	X	X	X
Diethylbenzene	monocyclic	X		X
9H-fluorene	polycyclic	X	X	X
Benzyl alcohol	monocyclic		X	
Pyrene	polynuclear		X	X
2,4-dimethylphenol	monocyclic		X	
Phenol	"	X	X	
2-methylphenol	"		X	
Dibenzofuran	polycyclic		X	
Bis (2-ethylhexyl) phthalate	monocyclic		X	

results did not represent PICs from VOST samples, however, while the SAIC results did. The Linak and Ryan et al.(5), results are not from bag burning but from simulated field burns of agricultural plastics used for mulching. Some of the Linak (5) PIC compounds were extracted from the residue after burning and did not appear in the emissions. There is a similarity among the compounds in all three studies however, in that many were mono- and polycyclic in nature.

Conclusions

Several conclusions may be drawn from the study thus far:

- Most PICs from the simulated open burning of Thimet pesticide container bags are polycyclic aromatic compounds. Cyclic compounds are produced from simulated open burning of agricultural product containers in both the laboratory and large, field-scale tests.
- More PICs are produced from burning used Thimet bags than from burning unused bags.
- PICs are produced from burning both used and unused bags.
- Naphthalene is a prominent PIC from burning used and unused bags. Compounds related to naphthalene are common among PICs.

Acknowledgment

The work performed was partially supported by the U.S. EPA, Office of Pesticide Programs. The work was executed within the scope of the U.S. EPA grant, No. CR 817460-01-0.

Literature Cited

1. *Executive Summary for Proposed 40 CFR Part 165*. U.S. EPA, May 1990.
2. *Pesticide Containers - A Report to Congress*. Red Border Draft, Office of Pesticide Programs, U.S. EPA, Washington, D.C., March 1991.
3. *The Bag Packaging Workshop Manual*. Stone Container Corporation, Bag Division. Undated, c. 1989.
4. Engleman, V.S., Jackson, T.W., Chapman, J.S.F., Evans, J.B., Martrano, R.J., and Levy, L.L. *Field Test of Open Burning of Pesticide Bags in Farm Fields*. Draft report to the U.S. EPA, Risk Reduction Engineering Laboratory, Cincinnati, Ohio, by Science Applications International Corp., August 1991.

5. Linak, W.P., Ryan, J.V., Perry, E., Williams, R.W., and DeMarini, D.W. *Chemical and Biological Characterization of Products of Incomplete Combustion from the Simulated Field Burning of Agricultural Plastics. JAPCA, v. 39,* No.6, June **1989**.

RECEIVED July 20, 1992

Chapter 7

Characterization of Emissions Formed from Open Burning of Pesticide Bags

D. A. Oberacker[1], P. C. Lin[1], G. M. Shaul[1], D. T. Ferguson[1], V. S. Engleman[2], T. W. Jackson[2], J. S. Chapman[2], J. D. Evans[2], R. J. Martrano[2], and Linda L. Evey[2]

[1]Risk Reduction Engineering Laboratory, U.S. Environmental Protection Agency, Cincinnati, OH 45268
[2]Science Applications International Corporation, San Diego, CA 92121

This report summarizes a study characterizing air emissions and residues from a common practice - open burning of used pesticide bags in farm fields. Two types of bags were tested - paper Thimet and plastic Atrazine bags, both with gram amounts of the original pesticides still remaining inside. Sampling and analysis performed during replicate burns generated extensive data on particulates, volatile, semi-, and non-volatile organics, dioxins/furans, metals and combustion gases. While the amounts of particulates were high, the toxic releases appeared small in terms of posing any significant health or environmental risk.

Summarized are the findings of a study of the farm practice of burning used insecticide and herbicide bags in open farm fields. This technique is used in some parts of the U.S. to dispose of used insecticide, herbicide, fungicide, and pesticide bags. The characterization of gaseous emissions, particulates, and remaining residues are the focus of this study. The study was sponsored by the United States Environmental Protection Agency Risk Reduction Engineering Laboratory (U.S. EPA/RREL), in conjunction with the Office of Pesticide Programs (OPP).

One insecticide and one herbicide were selected by EPA for this study. The insecticide was Thimet, O,O-diethyl S-(ethylthio)methyl phosphorodithioate, which has the formula:

$$
\begin{array}{l}
\qquad\qquad\qquad\qquad\quad S \\
\qquad\qquad\qquad\qquad\quad \| \\
CH_3-CH_2-O \\
\qquad\qquad\qquad\quad \searrow \\
\qquad\qquad\qquad\qquad\quad P-S-CH_2-S-CH_2-CH_3 \\
\qquad\qquad\qquad\quad \nearrow \\
CH_3-CH_2-O
\end{array}
$$

0097-6156/92/0510-0078$06.00/0

This compound is most commonly known as Phorate. Trade names for this insecticide include Agrimet, Geomet, Granutox, Rampart, Thimet, and Timet. The particular formulation used for this study was Thimet 20-G, a granular product consisting of 20% Thimet and 80% inert material. The inert material is primarily clay with a deactivator material to prevent the clay from acting as a catalyst to decompose the Thimet. The deactivator is commonly a glycol.

The herbicide was Atrazine, 2-chloro-4-ethylamino-6-isopropylamine-s-triazine, which has the formula:

```
                      Cl
                      |
                      C
                     / \\
 CH3               N     N
 |                 ||    |
 CH ---- NH --- C       C ---- NH ---- CH2 ---- CH3
 |                 \    //
 |                   N
 CH3
```

This herbicide compound, commonly known as Atrazine, has trade names including: AAtrex, AAtrex-Nine-O, Aktikon, Atazinax, Atranex, Atratol A, Candex, Cekuzina-T, Fenamin, Gesparim, Inakor, Primatol A, Primaze, Radazin, Vectal, Zeapos, and Zeazin. The particular formulation used for this study was a granular product containing 90% Atrazine, with the remaining 10% being inert material.

Conclusions and Recommendations

It is generally acknowledged by expert scientists and combustion engineers that open burning is an inefficient disposal method which potentially can generate and release both nuisance and hazardous air emissions and residues. Compared to the controlled burning in a hazardous waste incinerator, where destruction and removal efficiency (DRE) must be 99.99% or higher, open burning generally does not even achieve 99% DRE. Open burning can also represent a fire hazard.

Environmental concerns are heightened when the material that is open burned is something other than wood, paper, yard waste, etc. For example, open burning of certain plastics or other materials containing complex chemical compounds, such as pesticides, inks, and adhesives, as well as halogens and toxic metals, can increase environmental hazards.

Although collection and commercial high-temperature incineration is a technically sound alternative for the burning of used pesticide bags, it is much more expensive than open burning. Thus, open burning of pesticide bags is a widespread practice that has the potential to release hazardous pollutants into the environment. When the bags are open burned, factors such as the quantity and chemical makeup of both the bags and the residual pesticide can affect the completeness of burning and the degree of personnel and environmental hazard. Furthermore, the emissions and residues from the open burning of pesticide bags had never before been identified and quantified in any previous research project.

However, despite the potential hazards as outlined above, the actual types and quantities of pollutants measured in this study appear to represent relatively small releases of hazardous material. More work is needed to characterize the types and concentrations of pollutants that might result from open burning other formulations and types of pesticides and bags. Likewise, it is unknown how environmental hazards of pesticide release from open burning compares with the hazards from aerial pesticide spraying, and how environmental hazards of combustion emissions from burning of pesticide bags compares with the hazards from open burning of large farm fields treated with similar pesticides.

The major conclusions from this study were:

- This investigation was the first of its kind to determine the emissions from the open burning of used pesticide bags.

- The results should provide data for the U.S. EPA Office of Pesticide Programs in future regulatory decisions regarding packaging, disposal, and distribution systems for pesticides.

- Of the average 960 mg/bag of residual Thimet, the open burning of the bags released about 18 mg/bag (2%) of the Thimet to the atmosphere and left about 4.7 mg/bag (0.5%) in the solid residue.

- Acetone and benzene were the primary volatile organic compounds released to the air from the burning of used Thimet bags. Naphthalene and phenol were the primary semivolatile organic compounds released to the air.

- The particulate emissions from the combustion of Thimet bags were 4400 mg/bag. This amount is equivalent to 0.57 grains/dry standard cubic foot (dscf), which exceeds the regulatory limit of 0.08 grains/dscf for hazardous waste incinerators.

- Of the average 330 mg/bag of residual Atrazine, the open burning of the bags released about 42 mg/bag (13%) of the Atrazine to the atmosphere and left about 87 mg/bag (25%) in the solid residue.

- Acetone and benzene were the primary volatile organic compounds released to the air from the burning of used Atrazine bags. Naphthalene was the primary semivolatile organic compound released to the air.

- Chlorinated dioxins and furans were found at very low levels in the air emissions (less than 0.02 parts per trillion) and in the residue (less than one part per trillion). The levels were only slightly above those found in the laboratory blank.

- The particulate emissions from the combustion of Atrazine bags were 500 mg/bag. This amount was equivalent to 0.13 grains/dscf, which exceeds the regulatory limit for hazardous waste incinerators.

The major recommendations from this study are as follows:

- A survey of the major pesticide and container combinations and the methods of their disposal should be conducted. This survey should include quantities and frequencies of disposal.
- Additional pesticide types and their bags should be evaluated.
- The burning of a combination of two or more types of bags, or bags plus field wastes, should be tested in the same burning pile.
- Appropriate bioassay testing of the residues and particulates should be conducted.
- An analogous study related to the burning of large sugar cane fields, for example, should be undertaken.

Test Description

Field activities on this project were conducted in central Florida during the winter of 1991. The tests involved the burning, in triplicate, of used bags from Thimet and Atrazine application to crop areas. Clean (new, never filled) bags were also burned to provide a baseline comparison for the emissions. The Thimet bags consisted of four layers of paper and single layers of aluminum and polyethylene. The Atrazine bags were made exclusively from polyethylene. The Thimet bags, which had a capacity of 22.7 kilograms (50 lb) of Thimet when full, still contained, on the average, 4.8 grams (standard deviation 3.2 grams) of Thimet granules remaining. Since the granules contained 20% Thimet, the total amount of residual Thimet was 0.96 grams/bag. The Atrazine bag, which had a capacity of 11.3 kilograms (25 lb) when full, had, on the average, 0.37 grams (standard deviation 0.29 grams) of Atrazine granules remaining. Since the granules contained 90% Atrazine, the total amount of residual Atrazine was 0.33 grams/bag.

The tests were performed in a shed that had been adapted to simulate the conditions experienced during actual open burning of the used bags (Figure 1). A tray filled with sand lying flush with the floor of the shed was used as a simulated roadside surface for burning the bags. The inlet air was supplied by a wall-mounted fan, nominally rated at 1,000 standard cubic feet per minute to provide approximately one room air change per minute. A small fan was used to provide a 5 mph wind across the tray, typical of afternoon weather in central Florida.

Six bags were hand-loaded into the burn area and a rock was set on the top bag to keep the bags in place. The bottom-most bag was lit using a match or

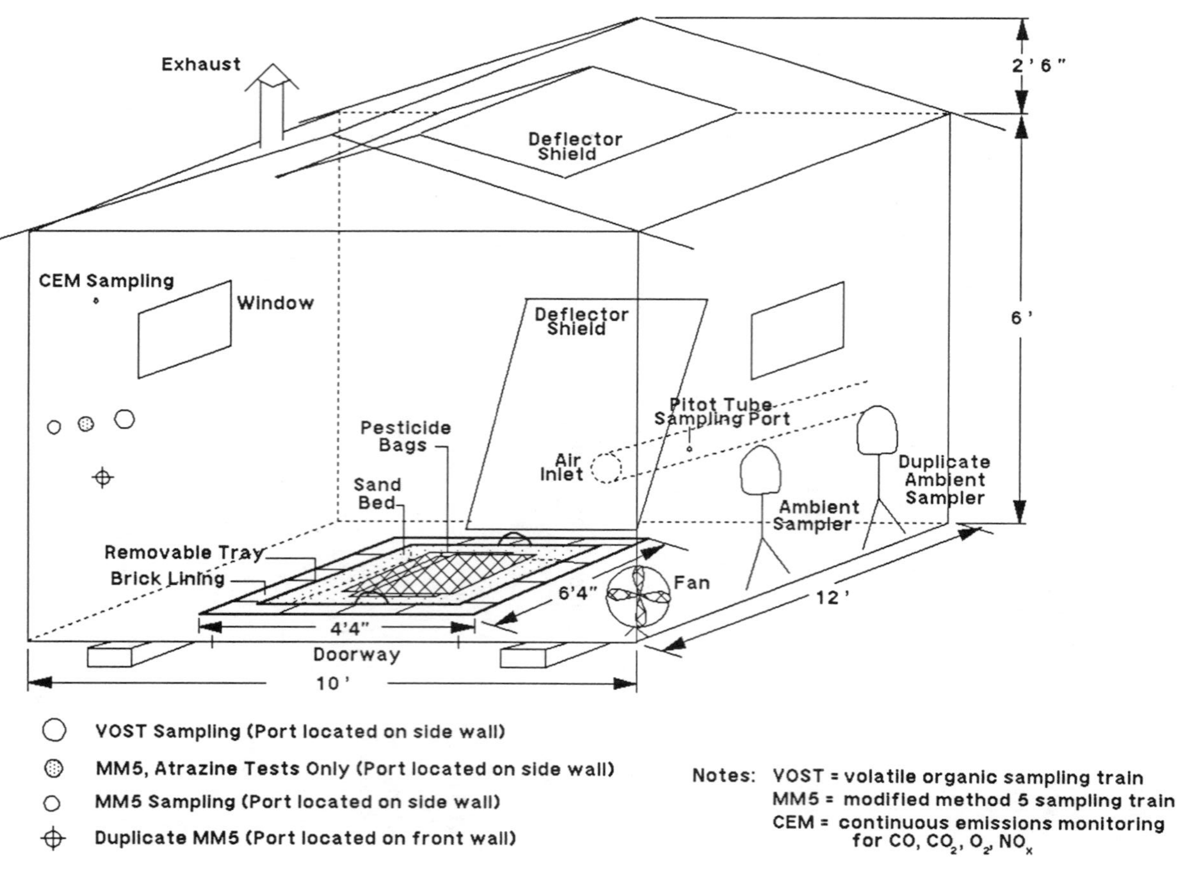

Figure 1. Shed simulation of open burning of pesticide bags.

rolled-up paper. The bags were allowed to burn until combustion slowed. Stirring or moving of the bags was required to enhance the burning, as in the case of the Thimet bags, or to push the unburned portions of the bags into the flame, as in the case of the Atrazine bags. Typically, the bags were stirred with a rake three times during a Thimet burn and stirred or moved between one to three times during an Atrazine burn.

Sampling of the air emissions inside of the shed were performed for volatile, semivolatile, and nonvolatile organic compounds, particulates, metals (Thimet tests only), and dioxins/furans (Atrazine tests only). Continuous Emission Monitors (CEMs) were used throughout the testing to determine concentrations of standard combustion gases and as a record of the behavior of the combustion process. The residue (ash and sand) left after the burning of the used bags was composited, aliquoted, and then analyzed for semivolatile compounds, including Thimet or Atrazine, metals (for the Thimet tests) and dioxins/furans (for the Atrazine tests). The leachability of the residue for semivolatiles and for metals was determined using the EPA Toxicity Characteristic Leaching Procedure (TCLP).

Results of Thimet Burns

To calculate the emissions from different numbers of bags, all the data generated from the field testing was reduced and reported on a per bag basis.

Air Emissions of Volatile Organic Compounds. The air emissions of volatile organic compounds from the combustion of Thimet bags are shown in Table I below. The major conclusions from the table are:

- Acetone, benzene, and toluene were the only targeted volatile organic compounds (as defined by SW-846 Method 8240) produced during the combustion of Thimet bags in quantities greater than those from the combustion of unused Thimet bags (baseline runs). Ethylbenzene, styrene, xylenes, and 2-butanone were produced in small quantities from the combustion of the bags themselves. Chloromethane and methylene chloride were also detected, but it is likely, based on field and laboratory blanks, that these are either ambient or laboratory contaminants.

- On average, the target compounds from EPA Method 8240 represented over 50% by weight of the total chromatographable volatile organic compounds (TCVO) emitted to the air. The other detected compounds (tentatively identified by gas chromatography / mass spectrometry) were primarily hydrocarbons.

- The target volatile organic compounds from the combustion of used Thimet bags, excluding methylene chloride and chloromethane, totaled 222 mg/bag. The target volatile organic compounds from the combustion of unused Thimet bags totaled 68 mg/bag. Subtracting the two, the increase in emissions of volatile organic compounds amounts to 16% by weight of the

Table I. Air Emissions of Volatile Organic Compounds Targeted in the QAPjP* Detected in the Baseline Run and Test Runs Burning Thimet Bags (Average Values in mg/bag)

Compounds	Background Run	Baseline Run	Test Runs
Acetone	1	14	63
Benzene	0.3	5	85
2-Butanone	ND	12	10
Chloromethane	2	1	7
Ethylbenzene	ND	5	5
Methylene Chloride	26	4	84
Styrene	ND	14	12
Toluene	2	7	36
Total Xylenes	1	11	11
Other Nontargeted Chromatographable Volatile Organics	8	65	243

ND - Compound was analyzed for but not detected.
* - Quality Assurance Project Plan

NOTE: Background Run = no burning;
Baseline Run = burning of clean bags

960 mg of input Thimet. Since the used bags were more weathered than the unused bags, these compounds could have resulted in part from poorer combustion of the bag material.

Air Emissions of Semivolatile and Nonvolatile Organic Compounds. The air emissions of semivolatile and nonvolatile organic compounds from the combustion of Thimet bags are shown in Table II. The major conclusions from the table are:

- Air emissions of semivolatile organic compounds were in the gas phase with very little on particulate.

- Air emissions of semivolatile compounds from burning used bags consisted primarily of Thimet and naphthalene with lesser quantities of phenol and 4-methylphenol. Similar quantities of naphthalene and phenol also were produced when burning unused bags.

- On average, the target compounds from EPA Method 8270 represented about 33% by weight of the total chromatographable semivolatile organic compounds (TCSO).

- The average Destruction and Removal Efficiency (DRE) for Thimet was 98.1%, where

$$DRE = \left(1 - \frac{WASTE\ OUT}{WASTE\ IN}\right)\ X\ 100\%$$

- The target semivolatile organic compounds represented about 9% by weight of the input Thimet.

- Air emissions of nonvolatile organic compounds were between 500 - 1000 mg/bag and are approximately equally divided between the gas and particulate phases.

Air Emissions of Total Particulates and Metals. The air emissions of particulates and metals from the combustion of Thimet bags are shown in Table III below. The major conclusions from the table are:

- There were no metals emitted to the air above the background levels. However, the background aluminum was anomalously high when compared to the total particulates, and therefore, the aluminum emissions of about 20 mg/bag from the baseline and test runs could have originated from the burning of the bag liner.

- The used insecticide bags emitted between 4000 - 5000 mg/bag of particulates. Burning six bags at a time produced the equivalent of 0.57 grains/dscf of particulates (corrected to 7% oxygen); this number exceeds the regulatory limit for hazardous waste incinerators of 0.08 grains/dscf at 7% oxygen equivalent.

Table II. Air Emissions of Semivolatile and Nonvolatile Organic Compounds Detected in Baseline Run and Test Runs Burning Thimet Bags (Average Values in mg/bag)

Compounds	On Particulate			In Gas Phase		
	Background	Baseline	Test	Background	Baseline	Test
Phenol	ND	ND	ND	0.2	8.4	13
Benzyl Alcohol	ND	ND	ND	ND	ND	Y
2-Methylphenol	ND	ND	ND	ND	ND	6
4-Methylphenol	ND	ND	ND	ND	3.7	10
2,4-Dimethylphenol	ND	ND	ND	ND	1.2 X	3
Benzoic Acid	ND	ND	ND	2.5	ND	Y
Naphthalene	ND	ND	ND	0.3	37	23
2-Methylnaphthalene	ND	ND	ND	ND	0.8	2
Acenaphthalene	ND	ND	ND	ND	1.2	3
Dibenzofuran	ND	ND	ND	ND	0.4	0.8
Diethylphthalate	ND	ND	ND	0.6	ND	Y
Fluorene	ND	ND	ND	ND	0.4	0.9
Phenanthrene	ND	ND	ND	ND	1.3	2
Di-n-butylphthalate	0.3 B	1.0 B	0.3 B	0.5 B	0.8 B	0.4 B
Fluoranthene	ND	ND	ND	ND	0.3	0.6
Pyrene	ND	ND	ND	ND	0.3	0.6
Bis(2-ethylhexyl)Phthalate	0.9	1.3	0.8	1.0	4.2	0.8
Thimet	ND	ND	ND	ND	ND	18
Total Chromatographable Semivolatile Organic	8.6	22	18	11	148	254
Total Nonvolatile Organic Compounds	ND	152	352	69	ND	274

ND - Compound was analyzed for but not detected.
Y - Compound was detected in only one of three runs.

Semivolatile Organic Compounds in the Residue. The semivolatile compounds in the residue from the combustion of Thimet bags are shown in Table IV. The major conclusions from the table are:

- The only targeted semivolatile compound from Method 8270 that was found in the residue was Thimet. Thimet represented about 25% by weight of the total chromatographable semivolatile compounds.

- The Thimet in the residue was found to be about 0.5% by weight of the Thimet in the used bags at the start of the test.

TCLP Results for the Residue. The residue from the combustion of Thimet bags was subjected to TCLP. The results are shown in Table V. The major conclusions from the table are:

- Small amounts of aluminum and barium were leachable from the remaining residue and surrounding soil as determined by TCLP protocols. The amounts were only slightly above the levels found in the sand blank.

- There were no detectable leachable organic compounds in the residue from burning used Thimet bags.

Results of Atrazine Bag Burns

As in the results of the Thimet bag burns, the results for the Atrazine bag burns have been reduced and reported on a per bag basis so that the data can be used to calculate the emissions from different numbers of bags.

Air Emissions of Volatile Organic Compounds. The air emissions of volatile organic compounds from the combustion of Atrazine bags are shown in Table VI. The major conclusions from the table are:

- Acetone, benzene, styrene, and toluene were the only targeted volatile organic compounds produced during the combustion of the used bags in quantities greater than those from the combustion of the unused bags. No other targeted compounds were produced at more than 2 mg/bag. During combustion of the unused bags themselves, chloromethane and methylene chloride were also detected, but it is likely that they were either ambient or laboratory contaminants. The emissions of volatile compounds during the combustion of used Atrazine bags were approximately one-third the weight of those emitted during combustion of the used Thimet bags.

- The target volatile organic compounds from the combustion of used Atrazine bags, excluding methylene chloride and chloromethane, total 71 mg/bag. The target volatile organic compounds from the combustion of unused Atrazine bags total 33 mg/bag. Subtracting the two, the increase in emissions of volatile organic compounds amounted to about 12% by weight of the input Atrazine.

Table III. Air Emissions of Total Particulates and Metals from Test Runs Burning Thimet Bags (Average Values in mg/bag)

Analyte	Background*	Baseline	Test
Total Particulate	66	1900	4400
Aluminum	90**	26	20
Antimony	ND	ND	ND
Arsenic	ND	ND	0.025 T
Barium	0.17	0.16	0.088
Beryllium	ND	ND	ND
Cadmium	ND	0.017	Y
Chromium	ND	ND	Y
Lead	0.042	0.053	0.063
Mercury	ND	ND	0.042 T
Silver	ND	ND	ND
Thallium	ND	ND	ND

* - No bags were burned in the background run. The entries are calculated as if 6 bags had been burned during the 30 minutes of sampling.
** - Aluminum value appears anomalous when compared to total particulate. Possible contamination.
ND - Compound was analyzed for but not detected.
T - Compound was detected in only two of three runs.
Y - Compound was detected in only one of three runs.

Table IV. Semivolatile Organic Compounds in the Residue from Thimet Runs

Analyte	Sand Blank	Baseline Run	Average Test Run
Thimet Concentration (mg/kg)	ND	ND	6.3
Thimet in Residue per Bag (mg/bag)	ND	ND	4.7
Total Chromatographable Semivolatile Organics (mg/bag)	ND	4.6	19

ND - Compound was analyzed for but not detected.

Table V. TCLP Results for Residue from Thimet Runs (Values Reported in mg/l)

	Sand Blank	Baseline Run	Average Test Run
Metal Analyte			
Aluminum	0.31	3.0	1.3
Arsenic	ND	ND	ND
Barium	0.067	0.53	0.13
Cadmium	ND	ND	ND
Chromium	ND	ND	ND
Lead	ND	ND	ND
Mercury	ND	ND	ND
Selenium	ND	ND	ND
Silver	ND	ND	ND
Semivolatile Analyte			
Thimet	ND	ND	ND
1,4-Dichlorobenzene	ND	ND	ND
2-Methylphenol	0.009 B	0.011 B	ND
3- and 4-Methylphenol	0.009 B	0.011 B	ND
Hexachloroethane	ND	ND	ND
Nitrobenzene	ND	ND	ND
Hexachlorobutadiene	ND	ND	ND
2,4,6-Trichlorophenol	ND	ND	ND
2,4,5-Trichlorophenol	ND	ND	ND
2,4-Dinitrotoluene	ND	ND	ND
Hexachlorobenzene	ND	ND	ND
Pentachlorophenol	ND	ND	ND

ND - Compound was analyzed for but not detected.

B - Compound was found in the extraction blank as well as the sample, indicating possible contamination.

Air Emissions of Semivolatile and Nonvolatile Organic Compounds. The air emissions of semivolatile and nonvolatile organic compounds from the combustion of Atrazine bags are shown in Table VII. The major conclusions from the table are:

- Air emissions of semivolatile organic compounds from the combustion of Atrazine bags were primarily in the gas phase with very little on the particulate.

- Atrazine and naphthalene were the primary semivolatile compounds being emitted to the air and were found only in the gas phase. Benzoic acid was detected in the test runs, but it was also detected in similar quantities in the background run when no bags were being burned.

- On average, the target compounds from EPA Method 8270, excluding benzoic acid, represented about 75% by weight of the total chromatographable semivolatile organic compounds. Atrazine alone represented about half of the total chromatographable semivolatile organic compounds.

- The average DRE for Atrazine was 87.3%.

- The target semivolatile organic compounds represented 15% to 20% by weight of the input Atrazine.

- Air emissions of nonvolatile organic compounds from the burning of used Atrazine bags were approximately 150 mg/bag and were approximately equally divided between the gas and particulate phases; these levels were one third to one sixth of those in the Thimet burns.

Air Emissions of Dioxins and Furans. The air emissions of dioxins and furans from the combustion of Atrazine bags are shown in Table VIII. The major conclusion from the table is:

- Dioxins and furans were found at very low levels in the air stream. The 2378-TCDD equivalence was only slightly above the levels found in the laboratory blank.

Air Emissions of Total Particulates. The air emissions of particulates from the combustion of Atrazine bags are shown in Table IX. The major conclusion from the table is:

- Particulates generated during the burning of the Atrazine bags ranged between 400 and 600 mg/bag. Burning six bags at a time produced the equivalent of 0.13 grains/dscf of particulates (corrected to 7% oxygen); this number exceeds the regulatory limit for hazardous incinerators of 0.08 grains/dscf at 7% oxygen equivalent. The particulate emissions from burning used bags were approximately equal to those from burning unused bags.

Table VI. Air Emissions of Volatile Organic Compounds Targeted in the QAPjP Detected in Baseline Run and Test Runs Burning Atrazine Bags (Average Values in mg/bag)

Compounds	Background Run	Baseline Run	Test Runs
Acetone	1	14	22
Benzene	0.3	12	22
2-Butanone	ND	2	3
Chloromethane	2	1	1
Ethylbenzene	ND	1	2
Methylene Chloride	26	3	22
Styrene	ND	2	9
Toluene	2	2	12
Total Xylenes	1	Y	1

Y - Compound was detected in only one of three runs.

ND - Compound was analyzed for but not detected.

Semivolatile Organic Compounds in the Residue. The semivolatile compounds in the residue from the combustion of Atrazine bags are shown in Table X. The major conclusions from the table are:

- The only targeted semivolatile compound from Method 8270 that was found in the residue was Atrazine. Atrazine represented about half of the total chromatographable semivolatile compounds in the residue.

- The Atrazine in the residue was found to be about 25% by weight of the Atrazine in the used bags at the start of the test.

Dioxins and Furans in the Residue. The dioxins and furans in the residue from the combustion of Atrazine bags were measured. The major conclusion was:

- Dioxins and furans were found at very low levels in the residue. The 2378-TCDD equivalence was only slightly above the levels found in the laboratory blank.

TCLP Results for the Residue. The residue from the combustion of Atrazine bags was subjected to the TCLP testing. The major conclusion was:

- The only leachable constituent of the residue in quantities above baseline and blank levels was Atrazine; the levels found were very low.

Combustion Efficiency of Open Burning of Pesticide Bags

The combustion efficiency for burning used pesticide bags was not as good as for a correctly designed incinerator. The combustion efficiencies for used Thimet and Atrazine bags (based on the CEM data for carbon dioxide and carbon monoxide) calculated from the formula,

$[CO_2 / (CO_2 + CO)] \times 100\%$,

Table VII. Air Emissions of Semivolatile and Nonvolatile Organic Compounds Detected in Baseline Run and Test Runs Burning Atrazine Bags (Average Values in mg/bag)

Compounds	On Particulate			In Gas Phase		
	Background	Baseline	Test	Background	Baseline	Test
Phenol	ND	ND	ND	0.2	0.8	2
4-Methylphenol	ND	ND	ND	ND	ND	0.3
Benzoic Acid	ND	ND	ND	2.5	ND	9
Naphthalene	ND	ND	ND	0.3	4.9	13
2-Methylnaphthalene	ND	ND	ND	ND	ND	1
Diethylphthalate	ND	ND	ND	0.6	0.3 X	0.3
Di-n-butylphthalate	0.3 B	0.5	Y	0.5 B	ND	Y
Bis(2-ethylhexyl)Phthalate	0.9	13 B	0.3 B	1.0	2.3 B	4 Y
Di-n-Octyl Phthalate	ND	ND	ND	ND	ND	Y
Atrazine	ND	ND	ND	ND	ND	42
Total Chromatographable Semivolatile Organic	8.6	13	1.7	11	29	89
Total Nonvolatile Organic Compounds	ND	483	68	69	NE	82

ND - Compound was analyzed for but not detected.
NE - No extract remained for the nonvolatile analysis.
Y - Compound was detected in only one of three runs.
X - Mass spectrum does not meet CLP criteria for confirmation but compound presence is strongly suspected.
B - Compound was found in the extraction blank as well as the sample, indicating possible contamination.

Table VIII. Air Emissions of Dioxins and Furans from Atrazine Runs (Values in ng/bag)

Analyte	Lab Blank	Run 1	Run 2	Run 3
2378-TCDD	ND	ND	ND	ND
2378-PeCDD	ND	ND	ND	ND
2378-HxCDD	ND	3	3	ND
2378-HpCDD	ND	6	4	ND
TOTAL TCDD	ND	0.4	ND	2
TOTAL PeCDD	ND	ND	ND	ND
TOTAL HxCDD	ND	4	4	ND
TOTAL HpCDD	ND	13	11	6.1
TOTAL OCDD	ND	ND	12	ND
2378-TCDF	2	ND	2	ND
12378-PeCDF	ND	ND	ND	ND
23478-PeCDF	ND	ND	2	ND
2378-HxCDF	ND	6	2	2
2378-HpCDF	ND	9	1	ND
TOTAL TCDF	1.5	2	3	ND
TOTAL PeCDF	ND	1	4	1
TOTAL HxCDF	ND	6	2	2
TOTAL HpCDF	ND	9	1	ND
TOTAL OCDF	ND	ND	ND	ND
2378-TCDD Equivalence (ng/bag)				
Method A	0.20	0.73	1.76	0.20
Method B	0.31	1.39	2.92	0.46
2378-TCDD Equivalence (parts per trillion)				
Method A	0.0013	0.0045	0.0112	0.0013
Method B	0.0019	0.0085	0.0186	0.0029

ND - Not detected.

NOTE: The 2378-TCDD equivalence was calculated using the scheme from I-TEFs/89 for Toxicity Equivalence Factors. Two methods were used for the calculation. Method A calculates the equivalence using a value of zero for any analyte that was not detected. Method B uses the value of the detection limit for that analyte in the calculation. Method A provides a lower limit for the equivalence, and Method B provides an upper limit.

Table IX. Particulate in Air Emissions from Atrazine Runs (Average Values in mg/bag)

Analyte	Background*	Baseline	Test
Total Particulate	66	630	500

* - No bags were burned in the background run. The entry is calculated as if 6 bags had been burned during the 30 minutes of sampling.

Table X. Semivolatile Organic Compounds in the Residue from Atrazine Bag Burns

Analyte	Sand Blank*	Baseline Run	Average Test Run
Atrazine Concentration (mg/kg)	ND	ND	76
Atrazine Residue per Bag (mg/bag)	ND	ND	87
Total Chromatographable Semivolatile Organics (mg/bag)	1.8	33	190

ND - Compound was analyzed for but not detected.
* - No bags burned. Quantitation reported is the calculated equivalent if the residue had been collected from a test burn.

was approximately 98% for both Thimet and Atrazine bags. Accounting for the unburned total hydrocarbons (THC), using the formula,

$$[CO_2 / (CO_2 + CO + 3THC)] \times 100\%,$$

where THC represents propane, the combustion efficiency dropped to 97% for both types of bags. The combustion efficiency of most well operated incinerators is well over 99%.

Full Project Report

For a complete presentation of all of the design, field testing, sampling, analytical methodology, detection limits, and quality assurance aspects including the entire array of test data resulting from this rather extensive field project, the reader is encouraged to obtain the full project report, prepared by SAIC under EPA Contract 68-C8-0061, Cincinnati Work Assignment 2-16.

RECEIVED August 25, 1992

Rinsate Minimization and Reuse

Chapter 8

Minimization and Reuse of Pesticide Rinsates

Ronald T. Noyes

Department of Agricultural Engineering, Oklahoma State University, Stillwater, OK 74078

Prior to 1985, pesticide waste management was not well defined. National EPA/NACA Pesticide Waste Management Workshops in 1985 and 1987 were held to identify, define and outline national pesticide waste handling, processing, storage and disposal problems. FIFRA '88 provided guidance for EPA, pesticide producers and users. University and chemical industry researchers have concentrated efforts on reducing pesticide rinsate and waste formation and disposal for both aerial and ground application. Developments of direct injection of full strength pesticides, expanded use of returnable containers, standardization of container openings, and improvements in container rinsing and closed mixing systems look encouraging. Direct oil and carbon filter/ozone processing of pesticide waste is being used by California applicators. A widespread/practical method currently used is reuse of rinsates as part of make-up water for subsequent applications or field rinsing and spraying back on the target area.

LEADING EDGE AND FUTURE MINIMIZATION TECHNOLOGY

Agricultural pesticide applicators today are faced with how to minimize pesticide waste. The best way -- ***don't create any!!*** New pesticide application technology -- direct injection at the nozzle, which keeps pesticides undiluted in modular closed loop set-on, lock-down recyclable/returnable containers, is being

0097–6156/92/0510–0096$06.00/0

marketed. In 1990, Deere and company introduced its new-concept "Lock & Load" planter-mounted applicator system for dry granular herbicides as a closed system. This development is setting the pace for liquid pesticide application.

At the EPA/TVA co-sponsored "International Workshop on Research in Pesticide Treatment/Disposal/Waste Minimization", February 26-27, 1991 at Cincinnati, Weeks (1991) outlined a new downstream injection system for sprayers and fertilizer applicators being developed jointly by Agway, Inc., Syracuse, NY and Raven Corp., Sioux Falls, IA. Agway's direct injection system embodies three refillable 15 gal pesticide tanks, stainless steel positive displacement metering pumps that deliver 5-200 ounces/min. These variable RPM, variable stroke pumps are interfaced with a radar ground speed computer for constant application rate control.

Currently, two U.S. companies, Raven Corp. and Midwest Technologies are producing direct injection field sprayer systems. Four other companies, Ag Chem Equipment Co., Lor-Al, Tyler and Willmar are offering direct injection as an option on their field applicators, (Schmuck, 1991). The quality and/or reliability of these injection systems have not been verified but conversations with chemical industry representatives indicate that additional research and development is needed to improve reliability and performance.

Other companies are expected to follow these leaders using 5-100 gal dedicated reusable/returnable containers that are part of interchangeable self-rinsing pump/hose/tank modules. Future U.S. systems may embody "smart" direct-injection variable volume spray applicators for pesticides and fertilizer based on computerized field data maps or on-board sensing. Computer research to develop variable field application rates for pesticides and fertilizers, conducted by Soviet university and mechanization institute engineering scientists was observed in the U.S.S.R. by Noyes (1991).

Closed loop pesticide handling systems have great potential for improving safety from pesticide poisoning for ground applicators. Eliminating carry-over pesticide field mixes and generating rinsates will save operating costs and reduce the potential safety hazard of applying the wrong pesticide on a sensitive target. This work should be advanced as rapidly as possible.

TODAY'S BEST RINSATE SOLUTION

The best management practice today with current technology aerial and ground sprayers is to minimize rinsate generation by field rinsing and spraying diluted field mixtures on target crops while the sprayer is still at the site, Figure 1. Aerial applicators do not require an FAA STC-authorized modification to add auxiliary rinse water tank systems with in-hopper pesticide containers. Currently, at least two companies market 20-23 gallon in-hopper rinsing systems for in-flight hopper and plumbing "triple-rinsing" that greatly improves pesticide application efficiency by spraying the rinsate on the target during the final trip. This minimizes or eliminates rinsate handling on mixing loading sites.

Figure 1. In-flight aircraft hopper rinse system. (Reproduced with permission from Curtis Smith.)

Ground applicators can modify sprayers with the available technology. When on-the-go field rinsing is not practical, the next best alternative is to rinse sprayers at mix/load sites, Figure 2, and transfer rinsates to 200-600 gallon high density cross-linked polyethylene, fiberglass or stainless steel holding tanks. Each tank should be marked for the individual pesticide. Poly or fiberglass tanks allow operators to see liquid levels in each tank and are usually cheaper than stainless tanks. Closed transfer systems using dry-break connectors are needed to minimize pesticide drips, reducing exposure of mixing/loading personnel and facility to field mix or rinsate pesticides.

Pesticides that are individually separated and identified, or that are in an identified, target compatible reusable mix are considered rinsates, not hazardous waste, by EPA. There's no limit on storage time of rinsates. However, its best to store only what you can use immediately. Rinsates can be used as make-up water for upcoming loads of identical field mixes to be sprayed on label registered target crops.

Field strength pesticides usually range from about 200:1 to 400:1 of shipping container strength. Rinsates are typically about 10:1 field strength dilutions. Each load ends up with some unused field mix. Normally about 3 to 6 gal of spray are left in the sprayer when pumps run dry at the end of a load. About 30 to 60 gal of water are used for rinsing the sprayer tanks and plumbing, depending on each sprayers size and configuration. Rinsate water/volume ratios of 10%/90%, 20%/80% or 30%/70% can be used to make up new sprayer loads. In a 20% mixture of rinsate that's been diluted during sprayer rinsing to 10% field strength, only 2% of additional active ingredient (AI) is added. Only 1% AI is added at 10% rinsate/90% water and 3% is added with 30% rinsate/70% water.

Most pesticide metering or measuring systems are less precise than this 101-103% range of accuracy. Keep in mind that the 1-3% AI in rinsate added back to the new mix is still part of the original calculated application if used on the same target, such as when sprayers are rinsed out daily, or the operator is waiting for a weather change to continue spraying.

Minimizing or eliminating storm water buildup on containment and/or loading pads will also reduce rinsate handling. Figure 3 illustrates minimum versus desirable open sided roof structures to shield pads from rainwater. Complete enclosures are the most positive means of complete stormwater elimination.

PESTICIDE WASTE TECHNOLOGY

Segregated, identified unused diluted pesticide field mixtures classified as rinsates can be stored indefinitely and can be legally used as outlined above. But what about mixed or combined rinsates that are not identifiable? EPA considers them as hazardous waste that must be legally disposed of within 90 days by a licensed hauler and a toxic waste facility.

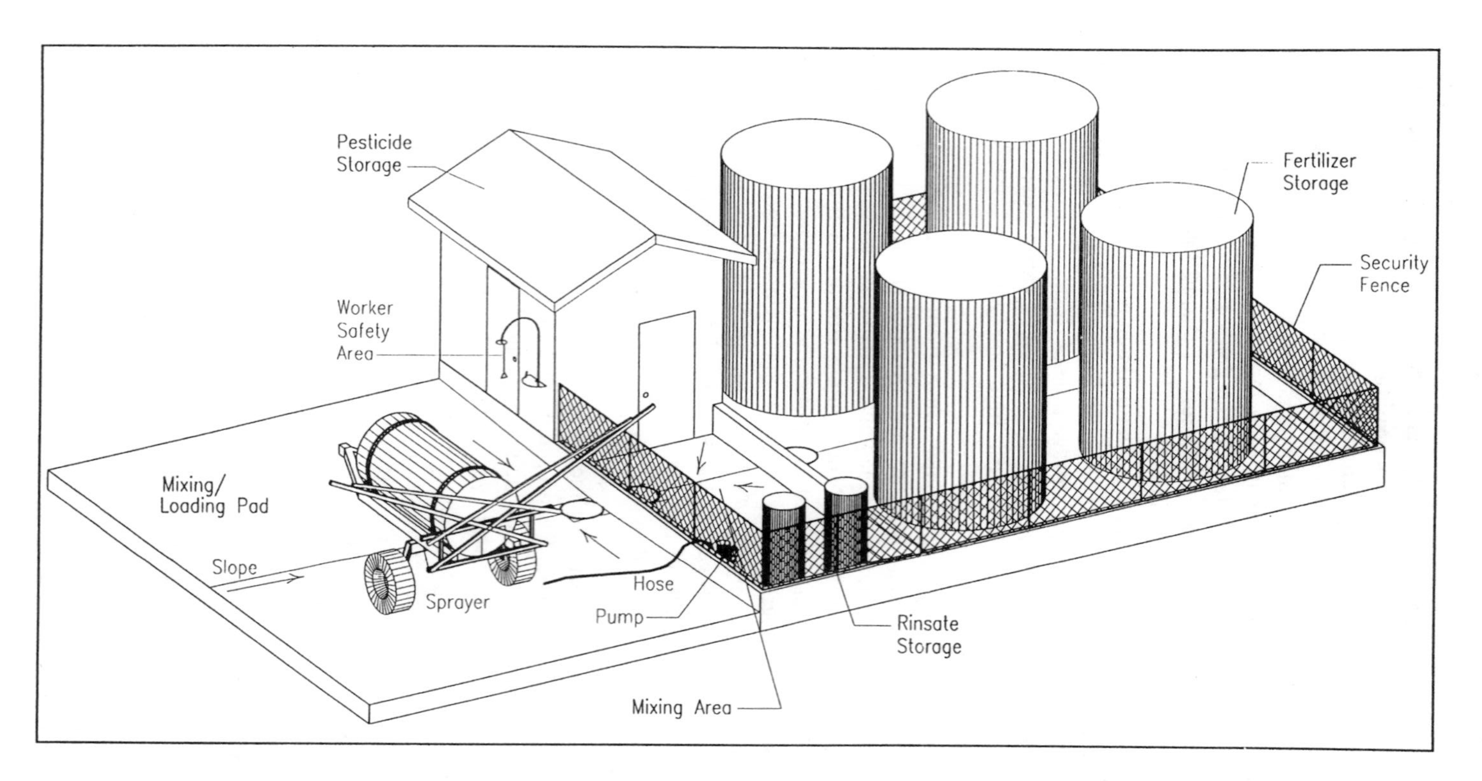

Figure 2. Medium sized multisump pesticide–liquid fertilizer mixing–loading facility. (Reproduced with permission from MWPS-37. Copyright 1992 MidWest Plan Service.)

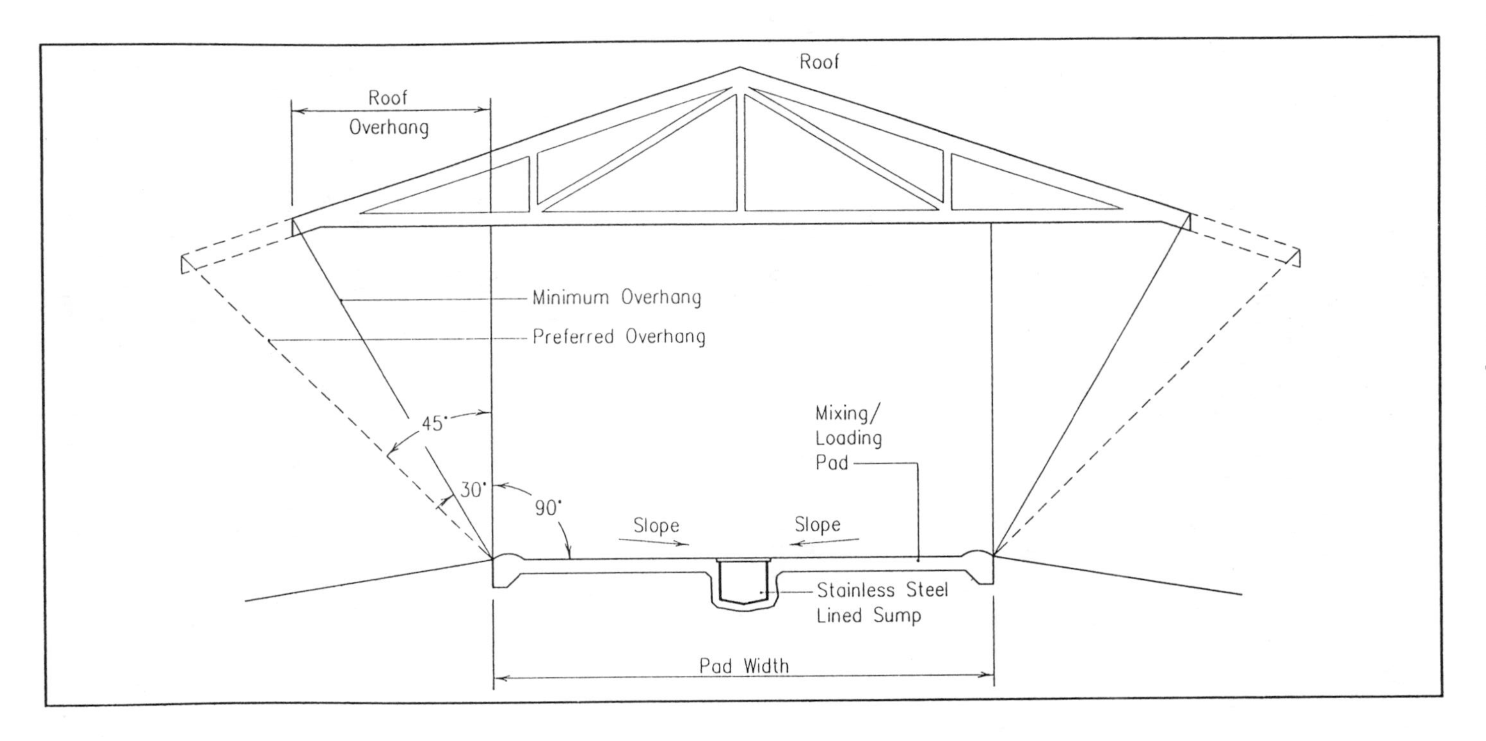

Figure 3. Open-sided roof structure over mixing–loading pad. (Reproduced with permission from MWPS-37. Copyright 1992 MidWest Plan Service.)

On-site treatment of pesticide waste has excellent potential for waste disposal. In 1990, at least four companies in the U.S. were marketing similar pesticide waste processing units. These equipment systems used a settling/ultraviolet-ozone degradation/filtering technology, Figure 4. Two of the systems added a flocculate in the settling chamber (initial hopper-bottomed waste holding tank) to increase sludge consolidation in the settling tank, reducing excessive filter buildup or plugging. Large particulate (paper), oil and activated carbon are used for in-line filtering. All filters require disposal (particulate, oil, and activated carbon) or high temperature regeneration (activated carbon).

These filtering systems are expensive processes. Are there operational problems? Yes. Words of wisdom and guidelines from meetings with ten California waste processing users interviewed in February 1990 were to: (1) Minimize total rinsate volume by field washing externally, in-field rinsing and spraying on target crops, and roofing facilities; (2) Use highly dilute solutions; (3) Use careful management. *They have already learned that waste processing is very expensive!!*

These waste processing sites were using activated carbon filtering systems installed by two California manufacturers to process pesticide rinsate waste. Some users found that trying to process highly concentrated mixtures created a problem of filter plugging or sealing off (bridging over near the filter inlet). They recommended diluting the rinsate to a low dilution-level field mixture to minimize the problem.

In addition to disposing or recharging the filters, disposal of sludge from the settling tank was a problem. It must be hauled by a licensed hazardous waste hauler to a hazardous waste disposal site. To minimize disposal costs, a unit-volume expense, minimizing liquid content of the tank-bottom sludge was important. One California operator drained the sludge into an open top chemical barrel and let it set in the sun to evaporate as much liquid as possible. A serious concern with that approach is possible health risks for unprotected or unaware workers due to breathing volatilized vapors from the exposed concentrated pesticide sludge.

One California university with one of the waste disposal systems was operating under the philosophy that "We're complying with California law, but we strongly encourage our people to avoid using the system when possible". They have their applicators field-rinse sprayers and spray the rinsate back on the target. The reason they're trying to minimize use is that they're trying to make their filters last as long as possible before they have to change them. They had not included money in their grant proposal budget for carbon filter recharging, sludge disposal, and disposal and replacement of the other filters. They were faced with a future expense which they couldn't afford.

So, even though this concept of pesticide waste disposal is on the market, there is substantial concern about its technical adequacy, and about the costs of

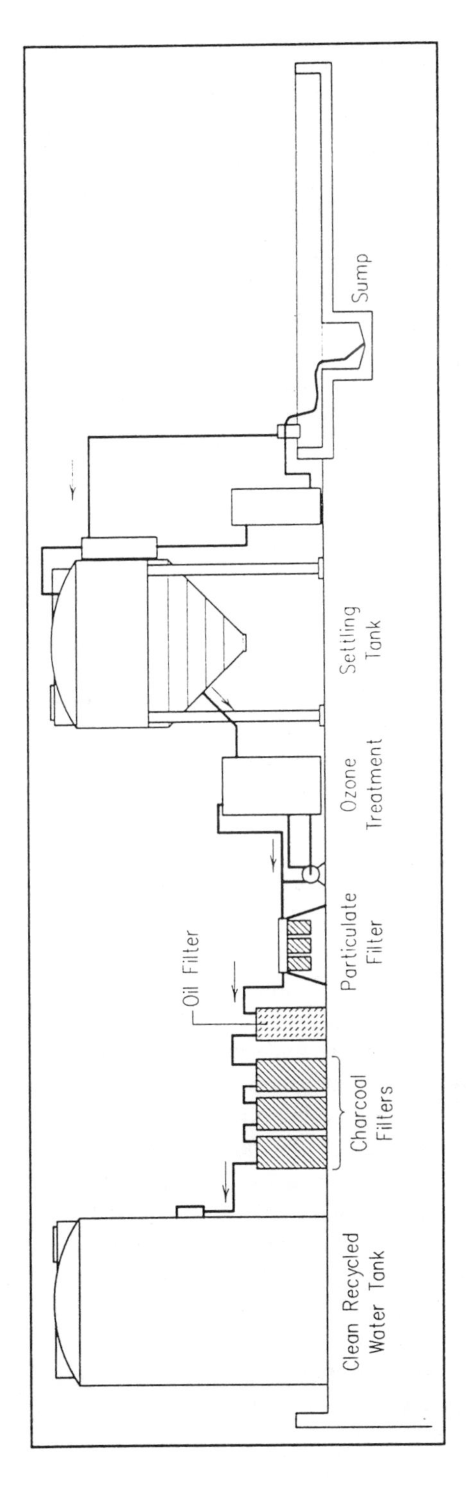

Figure 4. Carbon filter rinsate recycling system. (Reproduced with permission from MWPS-37. Copyright 1992 MidWest Plan Service.)

maintenance and operation, support service, and future replacement in the event that a superior, lower cost technology becomes available.

In its present form, these pesticide waste disposal systems are considered "closed loop" systems. The "clean" water stored after processing can not be used in mixing and spraying on fields. It cannot be released into surface drainage ways. **It's only use** is for re-use as sprayer rinse water to be recycled through the filtering and ozone cleaning system. This system concept presents a technological dichotomy. The rinsate waste must be diluted to a weak chemical-to-water ratio; the filtering/ozone system is physically limited to a specific hourly throughput flow rate, depending on the pesticide(s) concentration. During peak seasons, waste processing capacity may be a bottle neck when handling many pesticide products on several sensitive crops where thorough clean-out of sprayers on a daily basis is required. The solution under those conditions will probably involve adding enough rinsate waste holding tank volume and additional "clean water" holding volume to allow the waste filtering system to operate on a 24 hour/day continuous cycle.

RESEARCH

So far, we've discussed four pesticide rinsate and rinsate waste handling options and one method that eliminates field mixtures. *First*, direct injection and mixing at the nozzle, leaving full strength pesticide in returnable/reusable containers. *Second*, minimizing waste by field rinsing and spraying rinsate back on the target immediately. *Third*, segregating and storing rinsate for reuse as a dilute field-mix product, not as a waste product. *Fourth*, processing pesticide rinsate waste (mixed non-identified or field usable rinsates) that must be disposed of in less than 90 days to comply with RCRA regulations.

What work has been done or is in process that may provide future solutions to economical pesticide waste minimization or disposal? USDA researchers Kearney, Muldoon and Somich (1987) developed a twin tank rinsate disposal unit consisting of a two stage ozone and microbial degradation process. This system is relatively inexpensive and seems to have promising performance.

In a similar development, at Virginia Polytechnic Institute, entomologist Don Mullens et al. (1991) are focusing on concentration/containment methods using biodegradable nutrient-enriched lignocellulosic sorbents as a matrix where pesticides bond and degrade in composting environment. These researchers evaluated (1) steam exploded wood fibers, (2) activated carbon, (3) pine bark, and (4) peat moss absorbent composting media, for removal of several common pesticide compound formulations from aqueous suspensions using demulsification, sorption and filtration. Preliminary results from this work ***looks encouraging!***

Steve Dwinell (1991) of the Florida Department of Environmental Regulation has investigated an evaporation/degradation system for pesticide rinsate in a clear-roofed above-ground double-walled tank. Research results from this soil-filled absorption tank medium showed that the designed evaporation rates were more than adequate to keep pace with design rinsate handling rate. However, regulatory problems from a new interpretation of RCRA exemption for waste water treatment facilities required that treatment systems must have either a RCRA or NPDES permit. The future outlook of this process doesn't look optimistic. Research at North Dakota State, Figure 5, Michigan State, and Cornell Universities on similar system designs may run into similar obstacles.

TVA scientists Norwood and Gautney (1991) reported on research which involved removal of pesticides from aqueous solutions using liquid membrane emulsions. They indicated that additional research is needed before this concept is usable beyond research facilities. Gautney (1991) reported on some of TVA's research work which included batch oxidation of pesticides in rinsates, soil washing, solar evaporation/concentration/degradation of rinsate wastes, effects of best management practices on "natural remediation", and land application of rinsates. He reported that of these processes, only land application appears to be a recommendable practice today. Felsot and Dzantor (1991) discussed remediation of herbicide waste in soil as having potential for practical use based on their landfarming and biostimulation research.

TECHNOLOGY TRANSFER

Work that's closely related to pesticide rinsate storage and management, and pesticide waste processing involves development of facilities for pesticide mixing, loading and storage, and containment. Two modular watertight concrete mixing/loading containment facility designs are currently available from Oklahoma State University and the University of Wisconsin (Noyes and Kammel, 1989; Kammel and O'Neil, 1990; Noyes and Kammel, 1991). These designs form the core of a comprehensive Midwest Plan Service handbook, ***MWPS-37***, *"Designing Facilities for Pesticide and Fertilizer Containment"* (1992), co-sponsored by TVA. This handbook was written to provide EPA compliance guidelines on point-source groundwater protection at pesticide and/or liquid fertilizer facilities. Noyes designed large complex facilities with two or three sumps, Figure 6. A single-sump facility design, Figure 7, was developed by Kammel (1990) for small applicators/dealers. TVA agricultural engineer Mike Broder was instrumental in the MWPS-37 Handbook development by supporting this effort with travel funding so authors could meet periodically while developing the handbook.

The MWPS-37 Handbook development has been closely coordinated with EPA's OPP Division team during the development of the "Container Regulation." This regulation spells out groundwater point-source protection guidelines. The 116 page Handbook is intended to provide a baseline technology for state regulatory groups as they develop individual state environmental protection standards. Handbook development was coordinated with about 80 technical reviewers, including members of EPA's OPP Division,

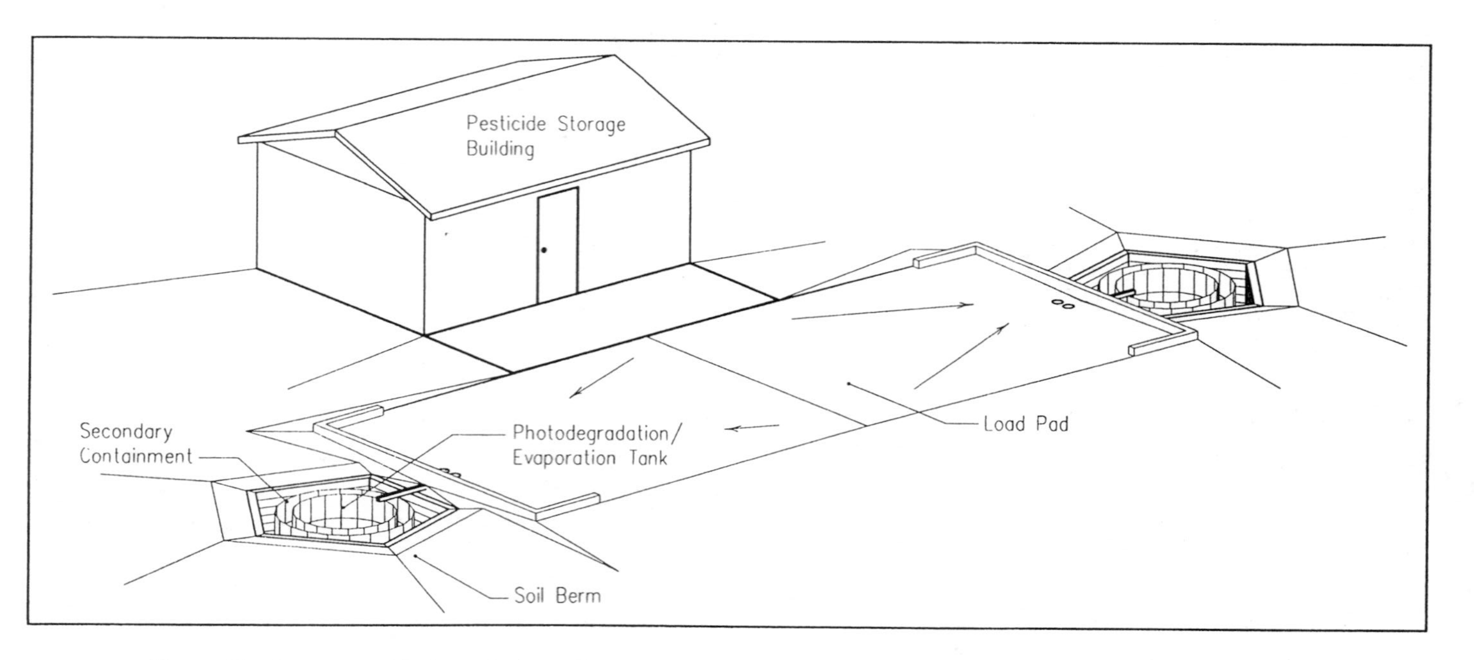

Figure 5. Research biodegradation disposal system. (Reproduced with permission from MWPS-37. Copyright 1992 MidWest Plan Service.)

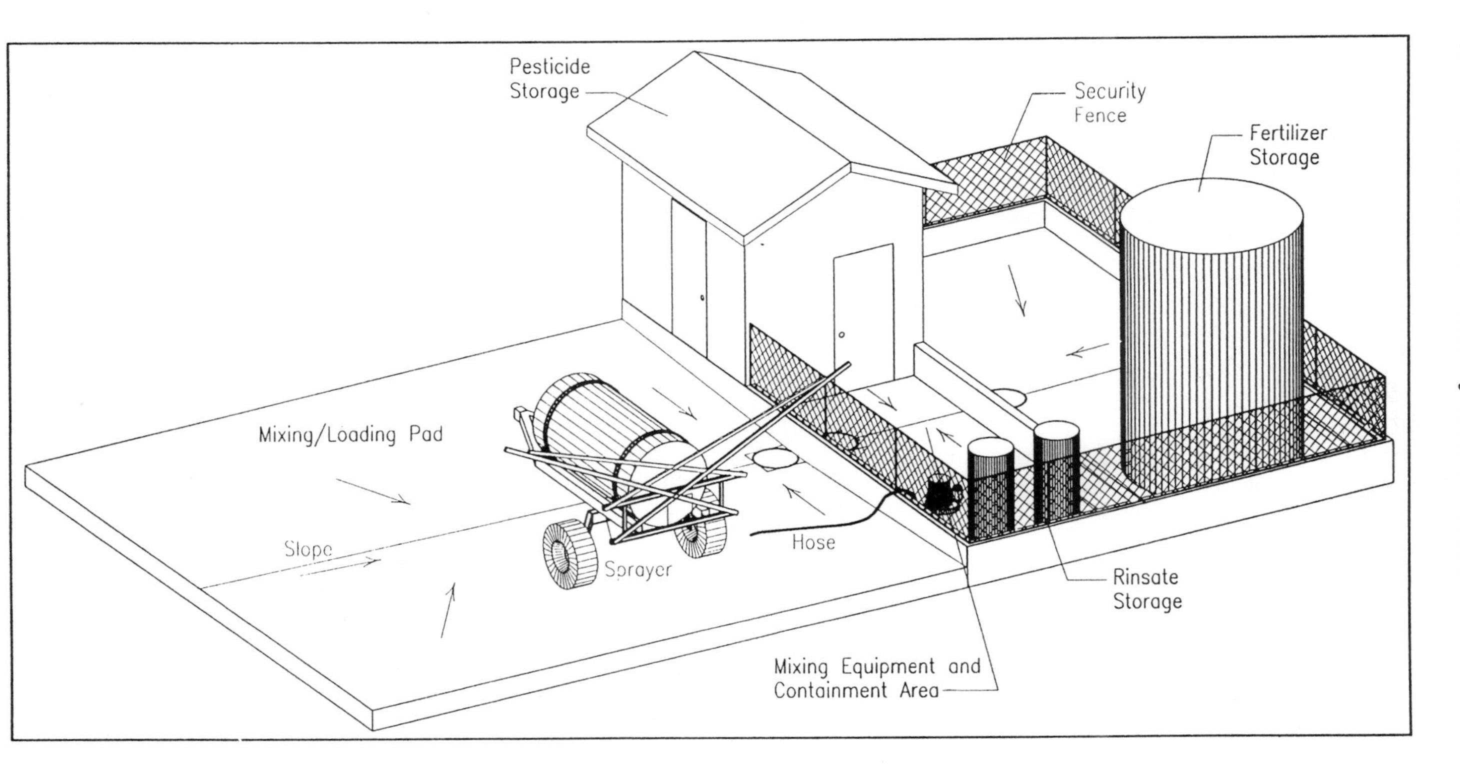

Figure 6. Large multisump pesticide–fertilizer containment pad. (Reproduced with permission from MWPS-37. Copyright 1992 MidWest Plan Service.)

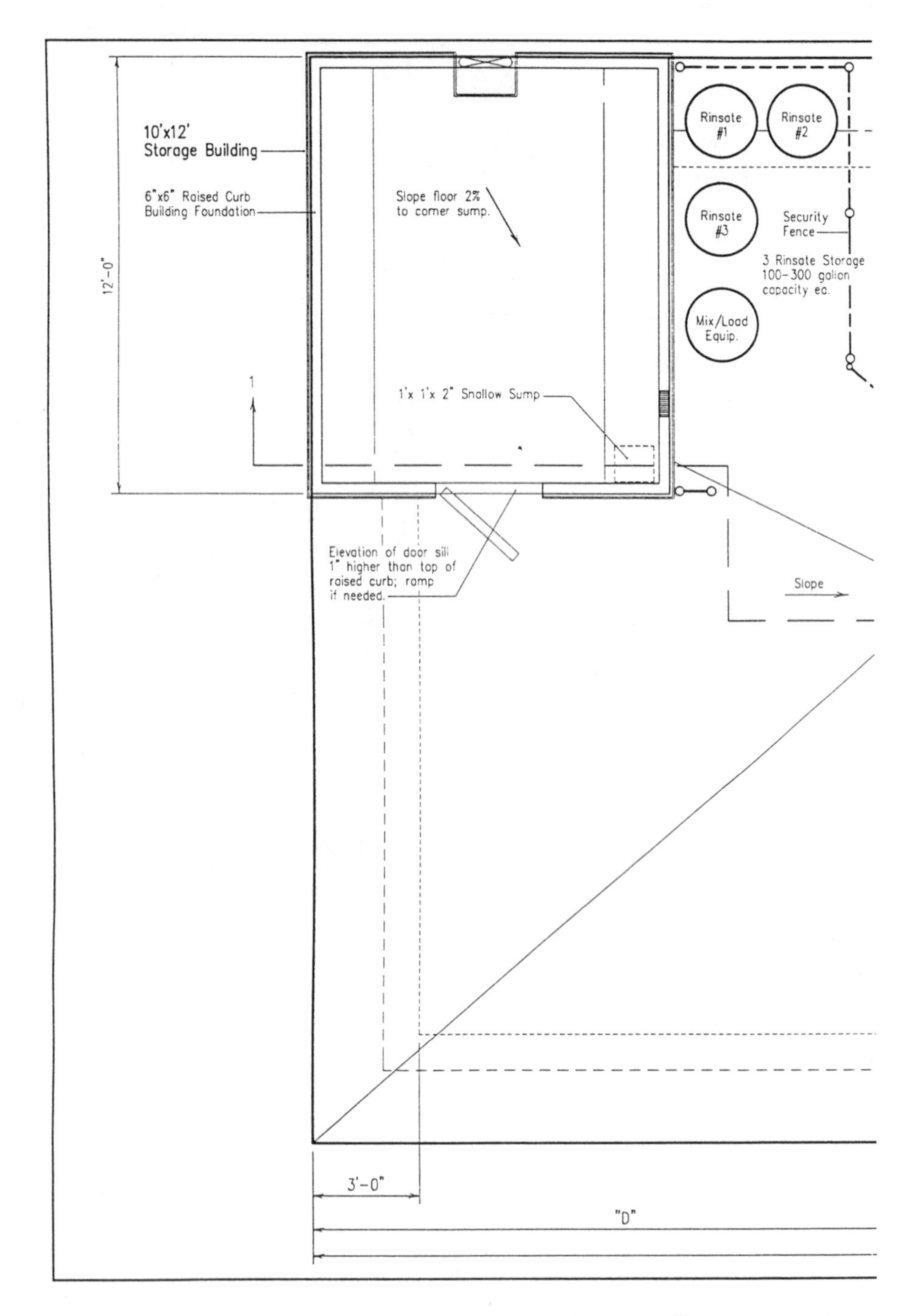

Figure 7. Plan view of single sump mixing–loading pad. (Reproduced with permission from MWPS-37. Copyright 1992 MidWest Plan Service.)

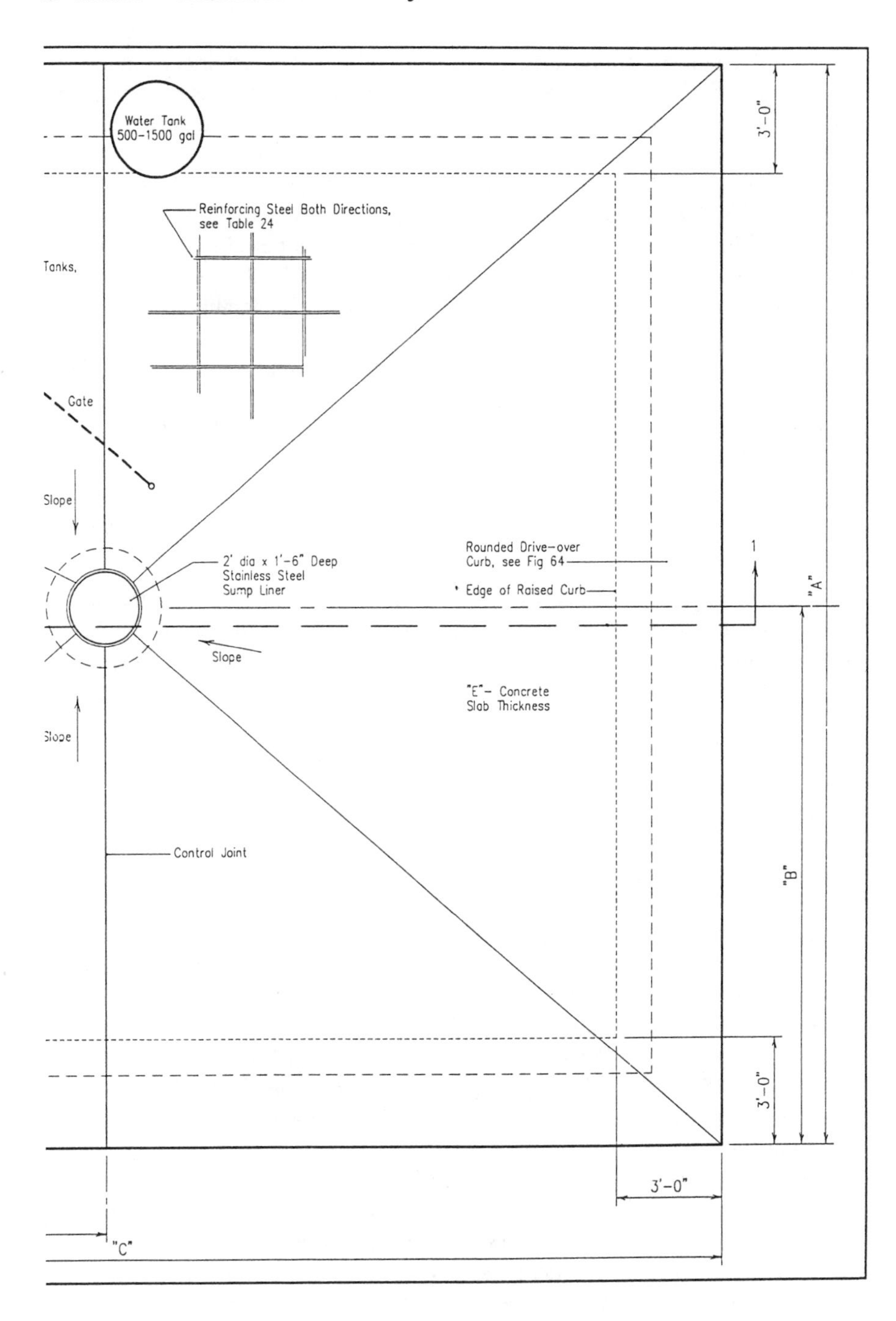
Water Tank
500-1500 gal
Reinforcing Steel Both Directions,
see Table 24
Tanks,
Gate
Slope
2' dia x 1'-6" Deep
Stainless Steel
Sump Liner
Rounded Drive-over
Curb, see Fig 64
Edge of Raised Curb
1
Slope
"E"- Concrete
Slab Thickness
Slope
Control Joint
3'-0"
"A"
"B"
3'-0"
3'-0"
"C"

agricultural chemical applicators, university and industry engineers, national chemical associations, and state regulators to provide a wide technical base of expertise for this agricultural chemical industry service handbook, scheduled for release in January, 1992.

As a way to enhance the initial use of this handbook, Midwest Plan Service (MWPS), Ames, IA is planning a pesticide/liquid fertilizer/ storage/handling/containment facility conference to be held in Kansas City in February, 1992. This conference will expand on the MWPS-37 Handbook to provide more immediate access to agricultural chemical facility containment technology. MWPS is also evaluating the development of plan sets, computer aided design packages of Handbook design drawings plus supplemental data sheets for designers. Plans would allow designers to rapidly expand the use of MWPS-37 Handbook design concepts for fast adaptation to local facility designs. These coordinated research and rapid technology transfer efforts will help get leading edge technology into use -- but, we need more.

FUTURE CONCERNS

There are major concerns in the agricultural chemical industry at all levels about economical control of environmental hazards and risks. Significant research efforts on pesticide waste minimization are being expended, but will they develop practical solutions soon enough? Are current research efforts adequate? Many industry people don't think so. Direct injection is the ideal process, but it is far from being a wide spread marketable product ready for use by many pesticide applicators. Is EPA working to help solve the problems as well as regulate them? Yes -- but this regulation/problem-solution process could be more productive if EPA (As well as NIH, NSF and others) supported research on critical topics was increased in concert with regulations. A major concern is that some states have moved ahead of EPA with regulations that are too stringent and not well coordinated across state lines.

More immediate dialogue by researchers as well as regulators, applicators and industry service support groups is needed to synthesize and harmonize a coordinated development effort. Glen Shaul, EPA Risk Reduction Engineering Laboratory, Cincinnati, 1991 workshop coordinator indicated that a 1992 follow-up EPA/TVA Research conference may be planned. He requested future workshop topics and informal research grant proposals be submitted during the 1991 workshop wrap-up.

Ag chemical dealers, manufacturers, applicators and other agricultural chemical industry professionals have reason to be concerned. It will take hard work and cooperation from all parties -- industry, EPA, university, the public, and Congress to solve the pesticide waste problem. Reasonable approaches to pesticide waste minimization can be accomplished if we communicate, support adequate research, develop technology transfer and keep working diligently on the problem until it's resolved.

LITERATURE CITED

Dwinell, S.E., 1991. "The Evaporation/Degradation System for Pesticide Equipment Rinsewater -- A Practical Solution to a Pesticide Waste Management Problem", Proceedings of the International Workshop on Research in Pesticide Treatment, Disposal, Waste Minimization, Cincinnati, OH, February 26-27, 1991.

Felsot, A. and Dzantor, K., 1991. "Remediation of Herbicide Waste in Soil: Experiences with Landfarming and Biostimulation", Proceedings of the International Workshop on Research in Pesticide Treatment, Disposal, Waste Minimization, Cincinnati, OH, February 26-27, 1991.

Gautney, J., 1991. "Tennessee Valley Authority, National Fertilizer and Environmental Research Center--An Overview", Proceedings of the International Workshop on Research in Pesticide Treatment, Disposal, Waste Minimization, Cincinnati, OH, February 26-27, 1991.

Kammel, D.W. and D. O'Neil, 1990. "Farm Sized Mixing/Loading Pad and Agrichemical Storage Facility", American Society of Agricultural Engineers Summer Meeting, Columbus, OH, June, 14p.

Kearney, P.C., Muldoon, M.T., and Somich, C.J., 1987. "A Simple System for Decomposing Pesticide Wastewater". Pesticide Degradation Laboratory, ARS, USDA, Beltsville, MD. Presented to Division of Environmental Chemistry, American Chemical Society, New Orleans, LA, August 30, 1987, 4p.

Mullens, D.A.., Young, R.W., Hetzel, G.H., and Berry, D.F., 1991. "Pesticide Disposal Using Demulsification, Sorption, Filtration and Chemical and Biological Degradation Strategy", Proceedings of the International Workshop on Research in Pesticide Treatment, Disposal, Waste Minimization, Cincinnati, OH, February 26-27, 1991.

MWPS-37, 1992. *"Designing Facilities for Pesticide and Fertilizer containment".* Kammel, D., Noyes, R., Hofman, V., Riskowski, G. 120 pages.

Norwood III, V.M., and Gautney, J., 1991. "Removal of Pesticides from Aqueous Solutions using Liquid Membrane Emulsions", Proceedings of the International Workshop on Research in Pesticide Treatment, Disposal, Waste Minimization, Cincinnati, OH, February 26-27, 1991.

Noyes, R.T. and Kammel, D.W., 1989. "Modular Concrete Wash, Containment Pad for Agricultural Chemicals". American Society of Agricultural Engineers Winter Meeting, New Orleans, LA, December, 32p.

Noyes, R.T. and Kammel, D.W. (1991). "Design Considerations and Criteria for Concrete Pads for Pesticide and Liquid Fertilizer Handling, Storage and Containment", American Society of Agricultural Engineers, Applied Engineering in Agriculture (Submitted to Transactions July 16, 1991.) 15p.

Schmuck, D. (1991). "Direct Injection Answers The Call", Farm Chemicals, 154 (8), August, 1991, p. 14-15.

Weeks, S. (1991). "Downstream Injection Equipment for Sprayers and Fertilizer Spreaders", Proceedings of the International Workshop on Research in Pesticide Treatment, Disposal, Waste Minimization, Cincinnati, OH, February 26-27, 1991.

RECEIVED May 18, 1992

Chapter 9

Treatment of Pesticide Wastes

Regulatory and Operational Requirements for Successful Treatment Systems

Steven E. Dwinell

Pesticides and Data Review Section, Florida Department of Environmental Regulation, Tallahassee, FL 32399–2400

Systems for the treatment of pesticide application equipment rinsewater must be designed with the pesticide user and with applicable regulations in mind. Treatment systems should be economically feasible to build and operate, have technology the pesticide applicator can understand and operate, produce a treated product that is less toxic and more degradable than the original rinsewater, and be in compliance with applicable regulations. Two systems available and in use in the United States, granulated activated carbon (GAC) filtration and evaporation/degradation systems, have these characteristics and are useful to pesticide applicators who can not manage their rinsewater in other, less expensive ways. The RCRA facility permit requirement for evaporation/degradation systems, however, limits the use of these systems, and removal of this barrier would increase the number of pesticide applicators managing pesticide rinsewater in ways that avoid environmental contamination.

Cleaning pesticide application equipment produces rinsewater that contains pesticide residues. Proper management of this pesticide rinsewater is necessary to avoid the contamination of soil, ground water, and surface water that can occur when this material is improperly discharged. Concentrations of pesticides in this rinsewater range from 1 to 1000 mg/l (*1*). Contamination of soil and water has been documented at a number of sites in the United States where pesticides have been improperly managed (*2*).

0097–6156/92/0510–0113$06.00/0

Management options for this rinsewater include re-application of the material as a dilute pesticide, re-use as a diluent for subsequent batches of pesticide, disposal as a waste, or treatment (*1, 3, and 4*). The first two options are the most widely used by pesticide applicators who are properly managing rinsewater. The last two, disposal as waste and treatment, are much less widely used because of the expense of these methods and the difficulties encountered in complying with the regulatory requirements that apply to these management methods.

Treatment is the application of a process that alters the chemical characteristics of the waste water to the extent that the rinsewater can be managed as a non-pesticidal or non-hazardous material. The process used may be physical, chemical, biological or a combination of these. This definition is more restrictive than that used in the Resource Conservation and Recovery Act (RCRA), which includes processes that alter the physical state or the volume of the waste material. The above definition is used here to make a distinction between processes that act only to reduce volume or alter the physical state of the rinsewater (such as simple evaporation) and those that alter the chemical characteristics and thus can render a waste non-pesticidal or non-hazardous.

Treatment systems that can be used successfully to manage pesticide rinsewater must have the following characteristics:

- Technology appropriate for the pesticide applicator
- Economic practicality
- Acceptable treatment capability
- In compliance with applicable regulatory requirements

Systems that lack any of these characteristics can not be successfully applied to the treatment of pesticide rinsewater. If the technology used is too complex, or requires constant monitoring or adjustment, the pesticide user - who produces the wastewater to be treated - will not be able or may choose not to use these systems. If the systems are too costly to construct, operate, or maintain, then, again, pesticide applicators will choose not to use them. The treatment of the rinsewater must, of course, be able to produce results that are environmentally acceptable. The treatment must result in products that are less toxic and more rapidly degradable than the original rinsewater. And finally, the systems must comply with applicable regulations. If they do not, then the systems can not be used legally. Illegal use can result in expensive fines and even criminal prosecution.

Researchers developing treatment systems must take these four characteristics into account if they hope to see their systems successfully used. Of these four, the regulatory requirements can be the most difficult to accommodate.

Regulatory Requirements

The basic regulatory requirements that operators and designers of pesticide rinsewater treatment systems must take into account are those imposed by two Federal laws -the Clean Water Act (CWA), and the Resource Conservation and Recovery Act (RCRA). These two laws, and the associated regulations promulgated and enforced by the United States Environmental Protection Agency (USEPA), establish certain requirements that treatment systems must take into account.

There are also requirements established by each state that may be more restrictive than the federal requirements. These include ground water protection laws, permitting requirements for non-domestic waste water treatment, and more restrictive definitions of pesticides regulated as hazardous wastes.

The CWA prohibits the discharges of pollutants to surface water bodies unless a permit has been issued under the authority of the Act. In order to obtain a permit, the discharged effluent must meet certain water quality standards. Discharges to a site where the effluent can enter surface water bodies through storm water run-off are included in these requirements. The water quality standards for pesticides are often very low concentrations -below 1 microgram per liter (ug/l). In Florida, for example, surface water quality standards have been established for seventeen pesticides, and all but three are below 1.0 ug/l. Endosulfan, a commonly used insecticide, has a water quality standard of 0.001 ug/l for waters used for recreation and wildlife - the category in which most surface water bodies are likely to classified.

The permit that is required for these discharges is issued through what is called the National Pollution Discharge Elimination System (NPDES), and will specify the amount of water that can be discharged, the location of the discharge, and the quality of the water that can be discharged. Discharging treated water without a permit can be an offense punishable by fines, and civil or criminal penalties. Treatment systems that produce a discharge must take the CWA into account if that discharge is to or could affect a surface water body. Discharges to land may be required to meet water quality standards under state ground water protection laws.

RCRA is the Federal law that regulates the disposal and management of waste materials in the United States. One portion of the law, Subtitle C, establishes requirements for materials that are identified as hazardous wastes. Under this law, certain pesticides, but not all pesticides, are considered hazardous wastes when the material is discarded as a waste. Many of the listed wastes are obsolete insecticides, but some are pesticides in common use today, for example - 2,4-D, methomyl, aldicarb, and phorate. O.R. Ehart (*5*) and F.W. Flechas (*6*) provide lists of those pesticides that are regulated as hazardous wastes under RCRA. Under RCRA, all wastes generated from the use of one of these pesticides are considered hazardous wastes. Moreover, addition of any amount of waste regulated as a hazardous waste to non-hazardous waste renders the entire waste amount a hazardous waste. This means that rinsewater produced when application equipment used to apply one of these pesticides is cleaned must be managed as hazardous waste <u>if it is discarded as a waste</u>. Mixing rinsewater containing pesticides regulated as a hazardous waste with other water renders the entire amount hazardous waste if discarded as a waste. Under the provisions of 40 CFR 261.2, treatment of the waste is considered discarding it as a waste, and, consequently, treatment of pesticides regulated as hazardous waste is subject to regulation under RCRA in many cases.

RCRA establishes a large set of requirements for the management of hazardous waste. Hazardous waste producers must report their activities to the USEPA and meet certain requirements for labeling of stored waste. Most importantly for pesticide rinsewater treatment systems, facilities that store, treat, or dispose of hazardous waste (TSD facilities) must obtain a RCRA facility permit. Facility permits are very difficult to obtain and expensive. In Florida for example, the cost for application fees alone are on the order of $7,000. These fees may be increased soon to $30,000 - $60,000. In addition to the fees for permit application, there are large costs associated with the necessary engineering studies, environmental monitoring and consultant fees necessary to win permit approval. The issuing agency may require that all waste management activities at the facility seeking the permit, including past activities , be brought into compliance with RCRA standards before the permit can be approved. This results in the need to conduct lengthy and costly site assessment and remediation activities before a permit can be issued. The net result of these requirements is that very few facility permits are ever granted. In Florida, the number of treatment facility permits granted for all industries is less than six. None have been granted for treatment of pesticide rinsewater. There is apparently only one facility in the United States that has a RCRA facility permit for treatment of pesticide rinsewater. This is a facility at Iowa State University in Ames, Iowa, and they report that the permit is very expensive to maintain (Sobottka, E., Iowa State University, personal communication, 1991).

For treatment of pesticide rinsewater, RCRA facility permits are not practical to obtain. The cost of the permit can exceed the cost of the treatment system by manyfold, and the required reporting and compliance inspections will be unattractive to most pesticide users. The bottomline for pesticide treatment systems is that the system has to be designed and operated in a manner that exempts the system from RCRA permitting requirements.

There are three ways for treatment systems to be exempt from RCRA treatment facility permit requirements:

1. There is no treatment of pesticides regulated as hazardous wastes under RCRA.
2. A permit is obtained under the CWA as a waste water treatment unit.

3. The treatment system is designed to be a part of a closed loop system.

If no pesticides regulated as hazardous waste under RCRA are treated by the system, then no RCRA facility permit is needed. This may be difficult to accomplish in some operations, since a number of commonly used pesticides are regulated under RCRA Subtitle C. With careful management of pesticide use, however, this approach may be possible.

If it is possible to obtain a permit for a treatment system under the CWA, then a RCRA facility permit can be avoided, even if RCRA regulated pesticides are treated. This is possible under the wastewater treatment unit exclusion, 40 CFR 261.3 (a) (2) (iv) and 40 CFR 264.1 (g) (6), which allows waste treatment subject to regulation under the CWA to be exempt from regulation under RCRA. Treatment systems that obtain CWA permits must, however, have discharges of treated water to a surface water body that meet the water quality standards established under that law.

The third option, a closed loop system, can take advantage of two possible exemptions from RCRA requirements. One is the exemption from RCRA permitting requirements for recycling operations. If the treated water is re-used in the treatment process, without discharge or disposal outside of the treatment system, the system may be considered a recycling system and the waste not considered a hazardous waste under 40 CFR 261.2 (e). Alternatively, as is the case in certain carbon filtration treatment systems discussed below, the re-use of the treated water may be considered to "not constitute disposal to land" and, as long as any spent filtration units are properly disposed of within 90 days, the provisions of 40 CFR 262.34 will be complied with and the system will not need a RCRA facility permit.

Existing Treatment Systems

The pesticide rinsewater treatment systems in use legally in the United States can be divided into two general classes - carbon filtration treatment systems and evaporation/degradation treatment systems. Treatment using systems that do not comply with existing regulatory requirements does occur, but the extent is unknown, and these treatment systems are not discussed here. Treatment systems under development are also not discussed here.

Carbon Filtration Treatment Systems

The most commonly used legal treatment systems for pesticide rinsewater in the United States are those that employ granulated activated carbon (GAC) to remove the pesticide residues from the rinsewater. The use of GAC filters to remove organic contaminants is widespread in a number of industries, and these systems represent an application of a well-known and understood technology to the pesticide rinsewater management problem.

There are no published figures available on numbers of these systems in use, but industry sources estimate about forty GAC filtration systems in use for pesticide rinsewater treatment.

GAC filtration systems function by the exposing of the pesticide contaminated rinsewater to carbon particles. The organic pesticides are adsorbed onto the carbon and thus removed from the rinsewater. Figure 1 is a simplified schematic of this process.

The exposure time and capacity of the carbon for adsorption are important factors in the design and operation of these systems. From an operational standpoint, the finite capacity of the carbon for adsorption is critical. The operator of the system must know when the capacity has been exceeded in order to replace the carbon and continue successful treatment. The capacity of the amount of carbon in any given system will depend on both the volume of rinsewater treated and the concentration of pesticide in the rinsewater. Simply put, the more concentrated the pesticide in the rinsewater, the lower the amount of rinsewater that can be successfully treated. Table 1 illustrates this relationship for a given volume of activated carbon.

It is not a simple matter to estimate the pesticide residue concentration in a given volume of rinsewater. Chemical analysis is expensive and ELISA-type tests are only available for certain pesticides. System users must have some means of estimating when the pesticide adsorption capacity of the carbon has been exceeded.

There are two treatment products to be managed in these systems - the treated rinsewater, which may still contain low concentrations of pesticide, and the exposed carbon, which now contains pesticide residues. Pesticide applicators who use these systems must take the management of these materials into account when evaluating their advantages and disadvantages for their operations.

There are a number of firms that produce GAC filtration systems for industrial use. Two that have targeted the treatment of pesticide rinsewater are the Wilbur-Ellis Company and Imperial Chemical Industries, Ltd. (ICI).

The Wilbur-Ellis Company system has a number of features that illustrate the modifications necessary to handle pesticide rinsewater solutions with their mixtures of different formulations, oil contaminants, and soil particles. Figure 2 is a schematic of the Wilbur-Ellis system that illustrates these features - a solid particle filter, a settling tank, an oil filter, and an ozonation chamber. An ultra-violet light has also been added to the system downstream of the carbon filters to further degrade any bacteria or organics.

These features act to increase the useful life of the three carbon filters used in the system. The solid particle filter and settling tank remove soil particles that can clog pores in the filters downstream. The oil filter removes oils and greases that interfere with the ozonation process and carbon adsorption. The ozonation unit exposes the rinsewater to ozone and oxidizes organics in the rinsewater as well as any bacteria, algae, or other organisms that may act to foul the carbon filters. Oxidation of the organic

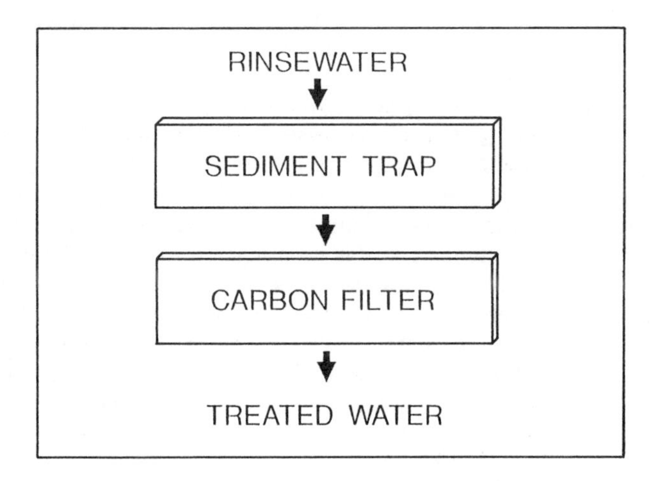

Figure 1. Schematic drawing of simplified carbon filtration system for treatment of pesticide rinsewater.

Table 1. Amount of waste water treated by 495 pounds of carbon as a function of contaminant concentration. Data courtesy of Wilbur Ellis Company

Contaminant Concentration (mg/l)	Carbon Loading (mg cont. / gram carbon)	Life of Carbon (hours)	Gallons of waste water treated
1	90.00	17793.06	5337918
50	291.03	1150.73	345218
100	358.30	708.73	212506
200	441.11	436.04	130813
400	543.08	268.42	80525
800	668.60	165.23	49567
1600	823.15	101.71	30513

pesticides in the rinsewater may also enhance their adsorption to the carbon. The carbon filters remove any residual ozone from the rinsewater. Figure 3 is a picture of a Wilbur-Ellis installation.

Currently, all spent filters and sludges generated through the use of the system are to be disposed of as hazardous wastes. The first carbon filter in the filter series is changed out after treating 50,000 gallons of rinsewater. The second filter is then rotated into the first position, the third into the second position, and a new filter placed into the third position. The water treated is stored for re-use in subsequent cleaning operations and then re-cycled through the system. There is no discharge from the system.

The ICI system is simpler and is designed to be a portable, modular unit. Pesticide rinsewater is treated in discrete batches of about 265 gals (1000 liters). Rinsewater is treated first with a flocculation agent and then put through a sand filter and two carbon filters. The carbon filters are supposed to be replaced after 20 batches or 5300 gallons. The flocculation chemicals added during the treatment process contain a dye that serves to indicate when the filters are no longer functioning properly. Sludge settling from the flocculation step and spent filters are to be disposed as solid waste and would have to be disposed of as hazardous waste if any pesticides regulated under RCRA Subtitle C were treated. The treated water can be discharged with a proper NPDES permit, or can be re-used in subsequent cleaning operations.

The GAC filtration systems described here have the characteristics necessary for successful treatment systems. The technology used is compatible with the abilities of pesticide users. Operators can keep track of the volume of rinsewater treated, and, as long as pesticide concentrations aren't unusually high, the volume guidelines recommended by the system designers should provide assurance that the carbon filters are changed before their adsorptive capacity is exceeded.

The systems are economically feasible for some applicators. Capital costs for these systems are $20 to $50 thousand dollars. Operational costs are primarily the costs of disposal of the filters and sludges, and these are on the order of $ 500 to $1,000 per disposal. These costs can be accommodated by some large pesticide application operations. These costs are much less than the cost of fines and clean-ups required if pesticide rinsewater is not properly managed.

The systems treat the rinsewater to produce a product (the effluent) that is less toxic and more rapidly degradable than the original rinsewater. The effluent has lower concentrations of toxic materials and degradation products. With sufficient treatment, it should be possible to meet water quality standards and obtain a discharge permit for the treated water. In most cases in the United States, however, the treated water is re-used as cleaning water for subsequent cleaning operations. The spent carbon and filters are managed as hazardous waste and there is thus no contaminated material released to the environment through this treatment process.

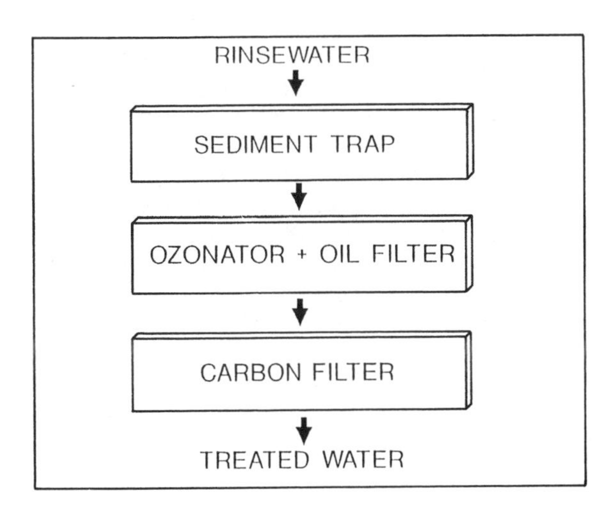

Figure 2. Schematic drawing of Wilbur-Ellis Company carbon filtration system for treatment of pesticide rinsewater.

Figure 3. Photograph of an installation of the Wilbur-Ellis Company carbon filtration system for treatment of pesticide rinsewater. (Reproduced with permission of Wilbur-Ellis Company).

Finally, the systems are in compliance with applicable regulations. Treated water is re-used to clean application equipment. Spent cartridges are disposed of as hazardous waste if any pesticides regulated as hazardous are treated. Since there is no disposal of the treated water, and if the spent carbon and other filters are removed within ninety days of becoming a waste, then the systems comply with RCRA regulations. In some cases the systems can be considered to comply with the provisions of RCRA that allow recycling systems to be exempt from permitting requirements. Wilbur-Ellis Corporation has expended a considerable effort to clarify the RCRA permitting status of their system, and RCRA regulators in the state and federal level have stated that these systems, when operated as closed-loop systems, do not need RCRA facility permits.

Evaporation/Degradation Systems

The other system type used to legally treat pesticide rinsewater in the United States is the evaporation/degradation system. These systems are of the type researched by Charles Hall and others in the 1970's at Iowa State University (7). The original systems consisted of a lined pit filled with a soil matrix into which rinsewater was placed. The liquid portion of the rinsewater evaporated and the pesticide residues were adsorbed onto the soil matrix and eventually degraded by micro-organisms in the soil. There was no discharge from the system.

The systems in use now have been modified to eliminate concerns about possible ground water contamination. These systems now typically use above-ground tanks (or double tanks if placement is below grade) to contain the matrix. A secondary containment system is also provided. The replacement of the pit with a tank allows the systems to be operated without the need for ground water monitoring. Leaks can be detected by inspection and corrected. The secondary containment systems provide extra assurance against ground water contamination should a leak occur. Figure 4 is a drawing of a system design used in Florida.

Evaporation/degradation systems are very simple to operate. Rinsewater is collected on a washdown slab for transfer to the tank or emptied directly into the tank. Solar radiation evaporates the water and pesticide residues are adsorbed to the soil matrix. Pesticides are degraded in the tank by bacteria or other mechanisms (such as hydrolysis or photolysis). There is no discharge of liquid from the system. The matrix in the tank is left undisturbed for the life of the system. When the system is dismantled, the matrix can be tested for residues and disposed of as a hazardous waste if necessary. In some cases, it may be possible to land spread the matrix if it does not exhibit hazardous characteristics. Depending on the durability of the construction materials, the system life should be fifteen to twenty years.

These systems treat rinsewater through the degradation of the pesticide component by microorganisms in the soil matrix. Pesticides are adsorbed by the soil matrix and may be only slowly degraded in these

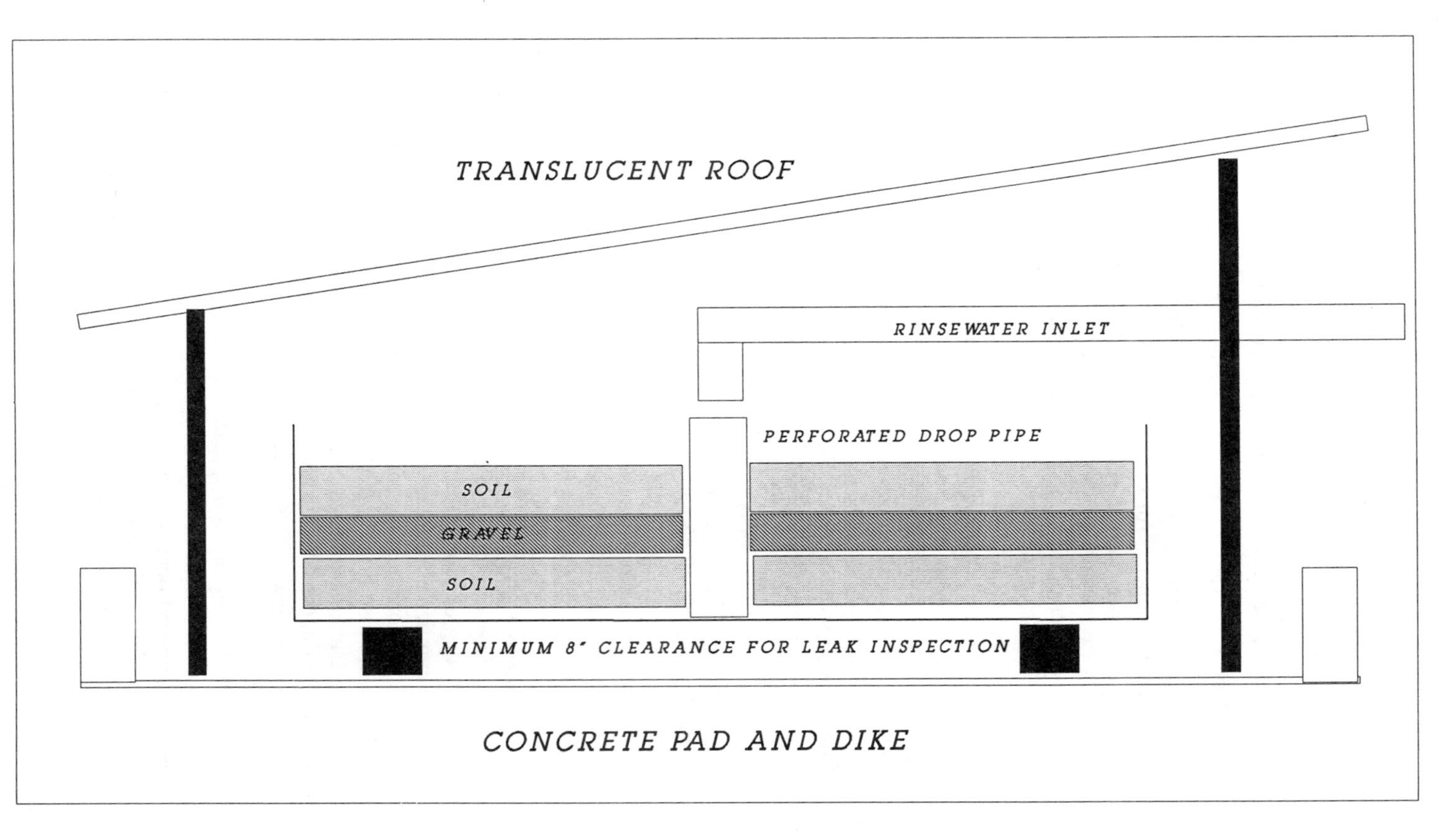

Figure 4. Drawing of evaporation/degradation system design used in Florida for treatment of pesticide rinsewater.

systems. The length of time required for degradation of a given pesticide may be long, but is not a concern during the operation of the system, due to the fact that the matrix is isolated from contact with ground water and there is no effluent that must meet a water quality standard before being discharged. Hall (*7*) and Hall et al. (*8*) provide data on the degradation rates of pesticides in these systems. It may be possible to enhance the degradation through the introduction of selected bacterial strains or nutrients.

One concern about these systems is air emissions of pesticides and pesticide degradation products. Air emissions from the early degradation pits at Iowa State University were measured by Hall et al. (*8*). Negligible amounts of pesticides were detected above the pits, with a median detection of 0.3 nanograms per liter of air sampled. Air emissions are expected to be low from these systems due to the fact that only dilute concentrations of pesticides are introduced. Concentrations are one to three orders of magnitude below the concentrations applied to target sites. Target sites are more likely to experience high concentrations of air emissions if a pesticide is associated with potential air emission problems.

The primary operational constraints for these systems is the capacity of the evaporation/degradation tank and the rate at which evaporation occurs. If the tank capacity is exceeded through poor management or precipitation, overflows will occur. Careful management of the amount of rinsewater generated and protection from rain and snow is needed. Storage capacity can be provided through the use of accumulation tanks for peak rinsewater generation periods. Evaporation can be enhanced through the use of clear roofing over the tank and setbacks from nearby buildings to allow for adequate airflow. By careful management of the amount of rinsewater introduced, overflows can be avoided.

Protection from rain and snow is most easily accomplished through the use of an adequate roof over the tank and any associated rinsewater collection pads. The University of Florida Agricultural Engineering Department recommends a roof with an overhang of thirty degrees measured from the edge of the pad to protect against blowing rain (*9*).

Evaporation/degradation systems are in use in a number of states, although the total number of systems is low, probably less than twenty. Systems are in use in Florida, Iowa, Michigan, and under construction in New York. There are also a number of older systems built using the original Iowa State University design in use around the country. Many of these still utilize the lined pit design and are probably in violation of state ground water protection laws. These systems may also be in violation of RCRA regulations.

The evaporation/degradation systems have the characteristics necessary for a successful treatment system. The technology is very simple, requiring very little operational skill on the part of the pesticide applicator. The primary responsibility of the operator is to not overfill the tank and to check for leaks. The systems are economically feasible, with capital costs in the $20,000 to $50,000 range and minimal operational costs. The disposal costs for the matrix are a one time cost that can be amortized over the life of the system. The treatment process results in products that are less toxic and more easily degradable than the rinsewater. The soil matrix should contain very little parent compound, and low concentrations of degradation products. The system has other environmental benefits since it requires very little energy and releases no waste products into the environment. For treatment of most pesticides, the systems are in compliance with existing regulations. Since there is no discharge, the system is not subject to the CWA. If no pesticides regulated as hazardous wastes are treated, then the treatment process is not subject to RCRA. If, however, pesticides regulated as hazardous wastes are treated, the evaporation/degradation system would have to obtain a RCRA facility permit under current RCRA interpretations.

This last requirement is the major drawback to the use of these systems for treatment of pesticide wastes. The systems currently in use either avoid treatment of pesticide rinsewater regulated as hazardous waste, or, in the case of the system in use at Iowa State University, have obtained a RCRA facility permit.

There is an effort underway at the state level to change this RCRA facility permit requirement. If the facility permit requirement can be eliminated, evaporation/degradation systems will have a wide application. It is in the interest of the USEPA to remove this requirement, since the use of evaporation/degradation systems by pesticide applicators would increase the options available to applicators for managing pesticide equipment rinsewater properly. Contamination of soil, ground water and surface water will thus be correspondingly reduced.

Conclusions

Proper management of pesticide rinsewater is needed in order to avoid contamination of soil, ground water, and surface water. Practical methods of pesticide rinsewater management are available that do not involve treatment. Re-application and re-use of the rinsewater as diluent are the two most widely used non-treatment options for applicators managing pesticide rinsewater properly. Treatment systems are needed by some applicators, however. In order to be successful, treatment systems should have the following characteristics:

- Technology appropriate for the pesticide applicator
- Economic practicality
- Acceptable treatment capability
- In compliance with applicable regulatory requirements

Carbon filtration and evaporation/degradation systems in use in the United States have the characteristics necessary for them to be useful to pesticide applicators who can not manage their rinsewater in other, less expensive ways. The RCRA facility permit requirement for evaporation/degradation systems, however, limits the use of these systems, and removal of this barrier would increase the number of pesticide applicators managing pesticide rinsewater in ways that avoid environmental contamination.

Literature Cited

1. Taylor, A.G.; D. Hanson; and D. Anderson. 1987. Recycling pesticide rinsewater. in *Proceedings: National Workshop on Pesticide Waste Disposal*. United States Environmental Protection Agency report no. 600/9-87/001.
2. Habecker, M.A. 1989. *Environmental Contamination at Wisconsin Pesticide Mixing/Loading Facilities: Case Study, Investigation and Remedial Action Evaluation*. September 1989. Wisconsin Department of Agriculture, Trade and Consumer Protection. Madison, Wisconsin.
3. Rester, D. 1987. Wastewater Recycling. in *Proceedings: National Workshop on Pesticide Waste Disposal*. United States Environmental Protection Agency report no. 600/9-87/001.
4. Dwinell, S.E. 1991. Managing Pesticide Wastes - What Applicators Can Do Now. in *Proceedings of the Environmentally Sound Agriculture Conference*, A.B. Bottcher, ed. University of Florida, Gainesville, Florida.
5. Ehart, O.R. 1985. Overview: Pesticide Wastes Disposal. in *Proceedings: National Workshop on Pesticide Waste Disposal*. United States Environmental Protection Agency report no. 600/9-85/030.
6. Flechas, F.W. 1987. Resource Conservation and Recovery Act Permitting of On-site Pesticide Waste Storage and Treatment. in *Proceedings: National Workshop on Pesticide Waste Disposal*. United States Environmental Protection Agency report no. 600/9-87/001.
7. Hall, C.V. 1984. Pesticide Waste Disposal in Agriculture. in *Treatment and Disposal of Pesticide Wastes*, R.F. Krueger and J.N. Sieber, eds. ACS Symposium Series no. 259.
8. Hall, C.V., J. Baker, P. Dahm, L. Freiburger, G. Gorder, L. Johnson, G. Junk, and F. Williams. 1981. *Safe Disposal Methods for Agricultural Pesticide Wastes*. United States Environmental Protection Agency report no. 600/2-81-074.
9. Bucklin, R.A.; D. Bottcher; and S. Dwinell. 1987. *Evaporation/Degradation System for Pesticide Equipment Rinse Water*. Bulletin 242. Institute of Food and Agricultural Sciences, University of Florida, Gainesville, Florida.

RECEIVED May 20, 1992

Chapter 10

Pesticide Application Systems for Reduction of Rinsate and Nontarget Contamination

Durham K. Giles

Department of Agricultural Engineering, University of California, Davis, CA 95616–5294

Rinsate and non-target contamination from agricultural pesticide application can be reduced through development of application technology which improves the efficiency of pesticide application. Reduced-volume application, coupled with auxiliary electrical or aerodynamic forces, decreases the amount of pesticide required for efficacious pest control through more efficient transport and deposition of spray droplets. Moreover, the required volume of tank mix and number of mix/load cycles is correspondingly reduced. Target-sensing sprayer controllers reduce the amount of pesticide applied through reduction of non-target deposition and contamination. Direct-injection sprayers eliminate tank mix and reduce rinsate through continuous, on-demand mixing of formulation and diluent. While each class of alternative pesticide application equipment offers attractive benefits, the design premise of each type also raises regulatory concerns over possible changes in potential worker exposure, environmental hazards and waste generation.

Creation of pesticide waste is a direct consequence of agricultural pesticide use; therefore, engineering developments which improve the physical and biological efficiency of pesticide use offer the potential for corresponding reduction of waste. Alternatively, engineering developments may be specifically targeted toward waste reduction.

Engineering Methods for Contamination and Waste Reduction.

Alternative Pest Control Strategies. The most direct route for reduction of pesticide contamination and waste is the development of non-chemical control methods for agricultural pests. Significant efforts are underway toward development of biological, cultural and mechanical techniques for insect, disease and weed control. An analysis of pest control alternatives for approximately 600 crop and pest situations common in California agriculture (*1*) was recently developed in response to potential cancellations of pesticide registrations (*2*) due to reregistration or other regulatory activity. Alternatives were found for 75% of the crop-pest combinations which would

0097–6156/92/0510–0127$06.00/0

be affected by loss of registered compounds. However, 60% of the alternatives were use of alternate compounds, rather than non-chemical techniques. Moreover, assessment of true availability of non-chemical alternatives was subjective and the study did not extend into economic analysis of the alternatives. While future developments may significantly reduce the need for agricultural pesticides, pesticide application will apparently remain important to production agriculture in the near future.

Improved Transport and Targeted Deposition of Pesticide. The typical agricultural spraying process generally results in spray deposition on less than 5% of the target foliar area with approximately 1% of the applied pesticide eventually being biologically active against the target pest (*3*). The potential for significant improvement in the atomization, transport and deposition of pesticide sprays has been recognized (*4*) as an engineering means by which the application rates of pesticide, and the corresponding waste problems, could be reduced. The principal methods for such improvements have been the delivery of pesticide in a more biologically active form, as in reduced-volume spraying, use of auxiliary forces for droplet transport and deposition, as in air-carrier and electrostatic spraying, and targeted deposition techniques, as in electronically-controlled spraying. Examples of these techniques are discussed in detail in following sections.

Engineering Developments Specifically for Reduction of Rinsate. In contrast to the above methods, which eliminate or reduce the amount of pesticide applied, engineering effort has also been directed toward development of improved systems for handling and preparing pesticide for application. Such approaches as in-line mixing, direct injection and closed mixing of concentrated pesticide with spray mix diluents specifically reduce the amount of pesticide waste or rinsate while using conventional atomization, transport and deposition techniques.

Implementation of Improved Application Systems.

While development of non-chemical pest control strategies represents an optimal solution for pesticide waste management, it will be considered beyond the scope of this symposium. The remaining two engineering approaches will be analyzed in detail.

Improved Transport, Deposition and Targeting of Pesticide. As previously discussed, the objective of this engineering development is the reduction in application rates of pesticide through more efficient application means. Often in agricultural pesticide application, the target area of the crop is a very small and specific site in relation to the land area being treated. Pesticide deposition on non-target areas can be considered waste, environmental contamination and a potential worker exposure hazard.

Reduced-Volume Pesticide Application. Prior to application, most pesticide formulations are mixed into water to form dilute "tank mixes" in which the diluent:formulation (mass) ratio can typically range from 50 to 800. Pesticide-use instructions included in label and registration documents typically specify the dilution ratio for various crops and application methods (aerial or ground-based). However, decades of study have found that biological efficacy of insecticides and fungicides generally increases with increasing concentration and decreasing droplet size (*3, 5-8*). In practice, the reduction of the diluent:formulation ratio, coupled with reduction in the size of the spray droplets has been typically called "low-volume" or "ultralow volume" spraying. However, the terms low and ultralow are relative and can be meaningless in consideration of the many combinations of pesticide formulations, crops and industry practices. The term, "reduced-volume" is preferable and succinctly indicates any practice using a diluent:formulation ratio less than that typically used for application of the particular pesticide on the particular crop.

In addition to the potential increase in biological efficacy and a corresponding decrease in active ingredient rate necessary for pest control, reduced-volume application results in an obvious reduction in the volume of tank mix handled during application. Moreover, tank size on reduced-volume equipment is generally smaller, requiring less wash water for cleaning and rinsing. The number of mix/load/refill cycles can be reduced, potentially reducing rinsate volume and applicator exposure.

However, simple reduction in droplet size and diluent:formulation ratio alone does not necessarily result in improved efficacy. Reduced-volume pesticide application is efficacious and appropriate only when the droplet transport and deposition processes are successful in achieving on-target pesticide deposit. The reduction in droplet size allows and, in fact, requires the addition of auxiliary forces for successful deposition.

The addition of aerodynamic forces for transport and deposition of small droplets has been common for over 30 years (*9*), received intense study (*10*) and become typical industry practice, particularly in orchard and vineyard culture. Hislop (*10*) concluded a review: "Since forced air currents are particularly suitable for transporting smaller spray droplets (ca. 40 to 150 μm), the use of this type of spectrum has led to economies in spray volumes, improvements in retention on targets and reductions in waste." Addition of electrical forces for improved deposition, i.e. "electrostatic spraying" is a common practice in non-agricultural spraying and has been extensively studied for agricultural use (*11*), albeit with limited commercial success.

Use of aerodynamic and electrical forces can not only increase the amount of spray material deposited but can also permit manipulation of the location of deposition. For example, pulsating air jets, tuned to the natural frequency of tomato flowers, have been used for removal, transport and deposition of pollen (*12*).

A current study (Giles, D.K., University of California-Davis, unpublished data) is investigating the target vs. non-target spray deposition from pesticide applications in greenhouse production of ornamental crops. When compared to a conventional high-pressure, high-volume (2300 l/ha) application, use of an electrostatic application (46 l/ha) and a "fogger" application (31 l/ha) achieved 4.8-fold and 5.4-fold increases in spray deposition on target foliage, respectively. Non-target deposition on the greenhouse bench surfaces was 3.8 times higher from the conventional application than from the reduced volume applications.

Electrical manipulation of target and non-target structures for altering the deposition of electrically-charged sprays has recently been investigated. Small fruit and vegetables are often grown on raised soil beds which are covered with plastic mulch film. The grounded plants protrude through the film and the film lies between the target plants and the soil surface. The film is essentially a non-target dielectric barrier underneath the target plants. Since the plants are earthed, their charge transfer relaxation time is brief in comparison to the dielectric film. Laboratory studies (*13, 14*) have confirmed that a pre-charging process can be used to differentiate the target plants from the underlying non-target mulch. By precharging the dielectric mulch with the same polarity as the charged spray droplets, deposition on the mulch film can be reduced. Correspondingly, the electric field created by the charged film enhances deposition on the target plants, particularly on the undersides of the targets.

A precharging system, coupled with a air-carrier electrostatic spray nozzle was used for laboratory evaluation of the selective deposition concept (*14*). When compared to uncharged spraying, use of spray charging and target manipulation reduced film deposition by 38% and increased upper and lower target surface deposition by 265% and 516%, respectively.

Target Sensing Sprayer Control Systems. Pesticide application has been described (*15*) as "the least efficient industrial process on earth." An efficient spraying system would deposit pesticide exclusively upon the desired biological target. Spray coating systems in non-agricultural applications with controlled environments and well-defined, non-variable targets can achieve near optimal performance. However, agricultural spraying is often done in adverse environments with highly variable target geometry. Often, the sprayer configuration and operating parameters are established for very general conditions and seldom altered for different crops, stages of growth or particular pest problems. As the spray target characteristics change due to crop development or simple variation within a field, application efficiency may significantly decrease. For example, in areas where target volume, area or mass is sparse, excessive pesticide may be released and deposited. Alternately, in areas of dense targets, poor spray deposition and biological efficacy may result.

The design premise of agricultural sprayer control is the continuous sensing of spray target characteristics and corresponding adjustment of the sprayer output as the targets are sprayed. The control process consists of three distinct phases, viz., sensing, decision and implementation. In the sensing phase, the presence, volume, density or similar characteristics of the spray target are detected. In the decision phase, the optimal amount of pesticide, location of release or other droplet transport characteristics are determined through algorithms based on the pesticide transport and deposition behavior of the sprayer, pesticide and the target. Finally, in the implementation phase, the control decision is physically implemented by variation of the sprayer output. Increased availability of electronic sensing and control components has resulted in development and commercialization of target-sensing sprayer control systems.

In the most elementary case, a single spray nozzle (or distinct collection of nozzles) may be controlled by the presence of a corresponding spray target. The nozzle is kept inactive until a target is detected; when a target is present, the nozzle is activated. Prototype systems have been developed which have used: spring steel trip wires to sense plants by direct contact; electrical probes in which a detection circuit was closed by contact between grounded plants and sensor probes; and, photoelectric sensors in which the plants interrupted an infared beam (*16*). Field tests of intermittent sprayers in cabbage, cauliflower and peppers have resulted in 24 to 51% reduction in the amount of applied insecticide with little or no reduction in pest control efficacy (*17, 18*). Early intermittent spraying systems were limited to simple detection of an object protruding into a target sensing area. Subsequent systems were developed which used light reflectance to discriminate between plant material and soil and selectively apply herbicide to weeds in fallow fields (*19*). The reduction is applied pesticide achieved by such intermittent systems is directly related to the ratio of projected target plant area to land area. Pesticide savings are greatest during early season spraying when plants are small and distinct gaps occur between plants. As the growing season progresses and the gaps are filled with plant foliage, pesticide savings are correspondingly reduced. A field study of iceberg lettuce growth (*20*) found the majority of pesticide applications were made when less than 50% of the land area was covered by crop foliage and some applications were made when only 5% of the land area was covered.

The logical extension beyond simple detection of the presence of a spray target is the sensing of the quantity of target foliage present. If fact, the current practice of basing pesticide dosage and diluent volumes on land area rather than spray target foliar area, volume or mass continues to be challenged by pest control researchers. Foliar area or leaf area index *(3)* and enclosed volume and density of target plants *(21, 22)* have been proposed as logical bases for pesticide rate determination. Orchard sprayer control systems which estimate the quantity of spray target present and regulate sprayer output have been developed and commercialized. One system estimates tree height by ultrasonically detecting the presence of tree foliage at various heights *(23)* while another system estimates tree height and volume by ultrasonic measurement of the tree projection outward toward the sprayer *(24, 25)*. Field tests of the measurement-based system resulted in 24 to 52% reduction in applied spray liquid with little or no reduction in foliar deposition.

Current research efforts are toward development of machine vision sensors which use image analysis to discriminate between plant species *(26)* or detect foliar characteristics *(27)*. Such sensory systems would allow detection of weeds within crop foliage, plant orientation or estimation of pest infestation or damage.

Target-sensing sprayer control systems can potentially reduce rinsate and non-target contamination through two means, viz., the overall reduction in applied pesticide and the specific reduction of deposition on non-target surfaces such as soil. However, most sprayer control systems do not utilize improved transport and deposition techniques. Rather, the systems have been designed for simple retrofit to existing sprayer equipment. While overall process efficiency is improved by control systems, the inefficiency in droplet formation, transport and deposition on target foliage remains.

Application Systems for Reduction of Tank Mix and Rinsate. Disposal of excess tank mix and rinsate from equipment cleaning has been identified as a significant environmental contamination and potential worker exposure concern. Considerable engineering effort has been devoted to development of "injection" systems which eliminate the tank mixing of pesticide and diluent. The fundamental concept in such systems is the separate storage of the diluent, usually water, the pesticides and the adjuvants. Rather than the conventional mixing of the tank mix constituents, the materials are continuously mixed on-demand, in small quantities, just prior to leaving the spray nozzle. The concept has been commercialized by a number of vendors (*28, 29*). While particular implementations may vary slightly in design, their operation is essentially as follows.

The spray mix diluent water is stored and pumped in essentially the conventional manner of typical sprayers. The sprayer pump, flow and pressure control devices and plumbing systems are virtually identical to non-injection spray systems. As the diluent water approaches the spray boom, the pesticides and adjuvants are mixed with the water to achieve the desired spray mix concentration. The mixing typically occurs in a mixing chamber assembly. The mixed solution is then routed to the spray boom where the spraying process is accomplished in the customary manner using conventional agricultural nozzles.

Since the application rate of the pesticide active ingredient is directly related to the injection rate of formulation into the mixing chamber, some commercial systems (*28-32*) have incorporated electronic control into the pesticide delivery systems. The most common development has been inclusion of a ground speed sensor and pesticide pump controller to maintain a desired application rate independently of ground speed variations. Other systems allow the operator to adjust application rate in response to visually observed variation in pest infestation or crop density.

When spraying is completed, water or a cleaning solution is pumped through the pesticide pump and delivery system. This small amount of virtual rinsate is mixed with diluent water in the mixing chamber and sprayed from the spray boom. A common design goal is the minimizing of components, piping and internal volume of the pesticide delivery system in order to further reduce the volume of rinsate necessary for cleaning. Early systems used small, individual tanks for storage of pesticide (*28-31*) while most recent systems (*32*) use the original pesticide container as a storage tank. The latter approach is preferable since it further reduces rinsate requirements and eliminates any unnecessary transfer of pesticide from the original containers. Direct-injection sprayers, coupled with recent development (*33*) and increasing use of returnable and mini- or micro-bulk containers, offer the potential for virtually waste- and exposure-free pesticide transport from the manufacturer to the injection pump on the sprayer.

While the concept of direct injection spraying is straightforward, practical implementation has been impeded by physical constraints and limitations of such systems. Unless a pre-mixing system is used, injection systems are limited to emulsifiable concentrate and soluble liquid pesticide formulations. Each constituent (multiple pesticides or adjuvants) of a desired spray mix must have an individual tank, piping, pumping and injector system. Pumping systems for metering of the concentrated pesticide formulations must be extremely accurate and precise for a wide range of physical properties of the pesticide liquids. The flow rates of at least two (diluent water and pesticide) liquid delivery systems must be accurately controlled in order to achieve the desired spray mix concentration.

The injection of concentrated pesticide into a mixing chamber downstream of the diluent pump and upstream of the boom creates design and operational difficulties. Since the diluent is supplied at a high pressure (above the boom pressure necessary for spray atomization), injection of the pesticide into the diluent requires high pressure flow from the metering pump. High pressure pumping of pesticide concentrate is considered a potential worker exposure and environmental contamination hazard and is prohibited by regulation in California (*34*). Use of a central mixing chamber also increases the response time of any injection controller system and reduces uniformity of the boom discharge. If the injection rate of the pesticide is altered, (e.g., in response to ground speed changes), the altered concentration of spray mix must exit the chamber and travel throughout the boom supply system. Considerable time can elapse as the concentration front travels through the boom; during the transient, the concentration of the spray mix from each nozzle can differ (*35, 36*).

An alternate design approach is to eliminate the mixing chamber and inject the concentrated pesticide upstream of the diluent pump. Such a design results in contamination of the high-volume pump and additional piping. Moreover, the sprayer cannot use bypass control for system pressure regulation since pesticide mix leaving the pump cannot be returned to the diluent storage tank. Moving the injection point further upstream increases the time delay between injection rate changes and the uniform discharge of the spray mix form the nozzles.

Other designs take a converse approach by eliminating the mixing chamber and moving the injection points further downstream, often to individual nozzle injection ports. The transient response time of the system is then greatly decreased. However, high pressure pumping of the pesticide is again required. Moreover, the high pressure pesticide supply system requires piping to each nozzle. With large spray booms and many nozzles, the internal volume of the pesticide piping system can become large and may contain a relatively high volume of pesticide in the line. Purging and rinsing of systems with long piping and multiple injection sites can require more time and generate more rinsate. Accurate metering of individual nozzle injection is difficult due to the extremely low volumetric flow rates of pesticide.

Injection spraying systems are only designed to reduce rinsate and contamination through improving the logistics of preparing the spray mix; spray atomization and transport is accomplished using conventional nozzles. In fact, most injection systems, like target-sensing controllers are designed as retrofit kits for existing sprayers. The injection mixing is not intended to improve spray deposition or biological efficacy of the pesticide application. The technology has not been considered a method through which significant reduction in pesticide application rates, non-target deposition or re-entry worker exposure could be achieved.

Regulatory Motivations and Constraints for Alternative Application Techniques.

Pesticide application is perhaps the most closely regulated agricultural activity. Subsequently, all engineering developments which alter the handling, application, deposition or persistence of the pesticide must consider the concomitant regulations and their intended effects. In addition to the direct economic considerations of pest control efficacy and effects on non-target organisms and the logistics and efficiency of the application, attention must be directed toward effects of alternative pesticide application techniques on regulatory constraints of application rate of active ingredient, volume of tank mix diluent, worker re-entry interval, pre-harvest interval and worker exposure.

Consider as a hypothetical example that a spray application technique is developed which completely eliminates spray drift and soil contamination by depositing all spray droplets on the target foliage. The sprayer is marketed and put into use by growers who continue to apply pesticides at full label rates. Since non-target deposition, which can be a significant proportion of the total pesticide applied, has been shifted to foliar deposition, the amount of pesticide on the target foliage would significantly increase. While such a machine would be environmentally attractive, the increased foliar deposition could create significant concern over re-entry worker or harvester exposure to higher dislodgeable foliar residues and food safety concerns over higher pesticide residue at harvest. Such concerns could be mitigated through mandated reduction in application rates to offset the increased efficiency of the application technique. However, the marketing of a sprayer system which required, rather than simply facilitated, a reduced application rate of pesticide could perhaps meet grower or industry resistance. The administrative burden of establishing and enforcing variable application (label) rates for different sprayer techniques could be significant. However, current federal regulation (*37*) allows "any method of application not prohibited by the labeling unless the labeling specifically states that the product may be applied only by the methods specified on the labeling."

Reduced-volume spraying, whether or not increased target deposition is achieved, is often limited or prohibited by regulation. FIFRA (*37*) prohibits application of pesticides at dilution (i.e., the diluent:formulation ratio) less than label specification. Accordingly, California regulation (*38*) prohibits "an increase in concentration of the mixture applied" unless it "corresponds with current published recommendations of the University of California." Since label specifications are typically developed through field testing using conventional application equipment, the labels are often written in such a manner as to exclude reduced-volume application technology.

Labels for pesticide use in ornamental greenhouse crops often state only a concentration and not an area-based application rate. The labels may specify a given amount of formulation to be added to typically 100 gallons of diluent water but may not specify how much mix should be applied per unit land or greenhouse area. The functional result is the prohibition of reduced-volume spraying but no limitation on the pesticide application rate.

The common concern with reduced-volume application and any technique which results in increased foliar deposition is the potential increase in worker exposure to pesticide from treated foliage. A field study and subsequent analysis of reduced-volume fungicide application in strawberries (*39*) found that increased deposition and longer persistence of pesticide could result in increased worker exposure. However, the effects could be mitigated by reducing the application rate of fungicide or extending the interval between applications.

The use of target-sensing sprayer control systems may also create regulatory rinsate concerns over the possible generation of excess tank mix. When an application rate is based on land area and the sprayer is accurately calibrated, a close estimate of the required amount of tank mix can be calculated for the area to be treated, resulting in a minimum of excess tank mix. When a control system is added in order to adjust the sprayer output for spray target density or intermittency, the applicator may not know *a priori* how much tank mix should be prepared. Uncertainty in the amount of required pesticide could often lead to generation of excess mix requiring disposal. The problem could be resolved by integrating direct injection pesticide mixing systems with sprayer control systems, eliminating excess tank mix while reducing non-target contamination.

Properly designed direct injection sprayer systems do not alter the deposition characteristics or active ingredient application rates of applied pesticide and are therefore relatively free of regulatory concerns. There can be, however, concern with the relatively high pressure pumping of concentrated pesticide and the potential for worker exposure. The concept and operation of direct injection systems is quite similar to those of closed mixing systems which are required in California. However, California guidelines (*34*) limit the pressure under which concentrated pesticide can be pumped to a maximum of 170 kPa. A typical direct injection system, with a nozzle boom pressure of 280 kPa would require pesticide to be injected at a pressure exceeding 170 kPa.

The regulatory concerns over alternative pesticide application equipment represent valid scientific questions regarding the characteristics of the systems and the subsequent effects on human and environmental non-targets. The concerns and the underlying questions are not intractable and their resolution should be considered part of the development process of alternative application technology.

Literature Cited.

1. Stimmann, M.W.; Ferguson, M.P. *Calif. Agric.* **1991**, *44*, 12-16.
2. Zalom, F.G.; Strand, J.F. *Calif. Agric.* **1991**, *44*, 16-20.
3. Hislop, E.C. *Aspects of Appl. Biol.* **1987**, *14*, 153-172.
4. Young, B.W. *Outlook on Agriculture* **1986**, *15*, 80-87.
5. Hall, F.R. *Aspects Appl. Biol.* **1987**, *14*, 245-256.
6. Hall, F.R. In *Safer insecticides - development and use;* Hodgson, E; Kuhr, R.J.; Eds. Marcel Dekker: New York, 1990; 453-508.
7. Himel, C.H. *J. Econ. Entomol.* **1969**, *62*, 919-925.
8. *Application and biology*; Southcombe, E.S.E., Ed. British Crop Protection Council Monograph No. 28; BCPC Publications: Croydon, 1985.
9. Potts, S.F. *Concentrated spray equipment*; Dorland Books: Caldwell, NJ, 1958. 598 pp.
10. Hislop, E.C. In *Air-assisted spraying in crop protection;* Monograph No. 46; British Crop Protection Council: Croydon, 1991; 3-14.
11. Law, S.E. In *Rational Pesticide Use*; Brent, K.J., Atkin, R.K., Eds.; Cambridge University Press: Cambridge, 1987; 81-105.
12. Nahir, D.; Gan-Mor, S.; Rylski, I.; Frankel, H. *Trans. ASAE* **1984**, *27*, 894-896.

13. Giles, D.K.; Law, S.E. *Trans. ASAE* **1990**, *33*, 2-7.
14. Giles, D.K.; Dai, Y.; Law, S.E. In *Electrostatics '91, Institute of Physics Conference Series*; O'Neil, B.C., Ed. Institute of Physics Publishing: London, **1991**, *118*, 33-38.
15. Rutherford, I. In: *Application and Biology*; Southcombe, E.S.E., Ed.; BCPC Monograph No. 28; British Crop Protection Council: Croydon, 1985.
16. Reichard, D.L.; Ladd, T.L. *Trans. ASAE* **1981**, *24*, 893-896.
17. Ladd, T.L.; Reichard, D.L.; Collins, D.L.; Buriff, C.R. *J. Econ. Entom.* **1978** *71*, 789-792.
18. Ladd, T.L.; Reichard, D.L. *J. Econ. Entom.* **1980**, *73*, 525-528.
19. Hooper, A.W.; Harries, G.O.; Ambler, B. *J. Agric. Eng. Res.* **1976**, *21*,
20. Giles, D.K.; Chaney, W.E.; Inman, J.W.; Steinke, W.E. *Trans. ASAE* **1991**, *34*, 367-372.
21. Sutton, T.B.; Unrath, C.R. *Plant Disease* **1984**, *68*, 480-484.
22. Giles, D.K. *J. Commercial Vehicles, SAE Transactions* **1989**, *98*, 257-265.
23. Roper, B.E. U.S. Patent No. 4,768,713; U.S. Dept. of Commerce: Washington, D.C., 1988.
24. Giles, D.K.; Delwiche, M.J.; Dodd, R.B. U.S. Patent No. 4,823,268; U.S. Dept. of Commerce: Washington, D.C., 1989.
25. Giles, D.K.; Delwiche, M.J.; Dodd, R.B. *J. Agric. Eng. Res.* **1989,** *43*, 271-289.
26. Franz, E.; Gebhardt, M.R.; Unklesbay, K.B. *Trans. ASAE* **1991**, *34*, 682-687.
27. Franz, E.; Gebhardt, M.R.; Unklesbay, K.B. *Trans. ASAE* **1991**, *34*, 673-681.
28. Landers, A.J. *Aspects Appl. Biol.* **1988**, *18*, 361-369.
29. Landers, A.J. In *Agricultural Engineering - Proceedings of the 11th International Congress*; Dodd, V.A.; Grace, P., Ed. A.A. Balkema Publishers: Brookfield, VT, 1989; Vol. 1; 2101-2110.
30. *Ag-Chemical Injector Model 240*; Weins, E.H.; Coleman, L.R., Ed. Evaluation Report 491; Prairie Agricultural Machinery Institute: Humbolt, Saskatchewan, 1986.
31. *Computorspray Spot Spraying Chemical Injection Metering System*; Atkins, R.P.; Russell, J., Ed. Evaluation Report 537; Prairie Agricultural Machinery Institute: Humbolt, Saskatchewan, 1987.
32. Landers, A.J. *Pesticide Outlook* **1989**, *1*, 27-30.
33. Mosher, P. *The Grower* **24**, *8*, 14, 27-30.
34. Rutz, R.; Gibbons, D. *Pesticide Information Series A-3* **1988**, California Department of Food and Agriculture, Sacramento, CA.
35. Tompkins, F.D.; Mote, C.R.; Howard, K.D.; Allison, J.S. In *Pesticide Formulations and Application Systems*; Chasin, D.G.; Bode, L.E., Ed.; ASTM STP 1112; American Society of Testing and Materials: Philadelphia, PA, 1990, Vol. 11.
36. Tompkins, F.D.; Howard, K.D.; Mote, C.R.; Freeland, R.S. *Trans. ASAE* **1990**, *33*, 737-743.
37. Federal Insecticide, Fungicide and Rodenticide Act as amended 1988, Sec. 2 (ee).
38. California Code Title 3, Div. 6, Chpt. 1. p. 359. 1991.
39. Giles, D.K.; Blewett, T.C. *J. Agric. Food Chem.* **1991,** *39*, 1646-1651.

RECEIVED February 14, 1992

Current Disposal Technologies

Chapter 11

Current Technologies for Pesticide Waste Disposal

James N. Seiber[1]

Department of Environmental Toxicology, University of California, Davis, CA 95616

Although much progress has been made in the past ten years, the problem of disposing of pesticide wastes, including rinsate from application equipment and containers, continues. Major progress has been made in minimizing the waste needing disposal, largely as a result of more thought and care by the pesticide manufacturers, formulators, and applicators. Research has also uncovered new and improved methods for physical, chemical, and biological treatment of wastewaters, some of which have achieved commercial utility. This paper will highlight development along the 15-year path leading to the current status of disposal technologies.

The pesticide waste disposal problem has come a long way towards practical solutions in the approximately 15 years since its existence began to attract attention. In the early 1970s it was difficult to generate much enthusiasm—much less practical research and development—perhaps because pesticide containers, wastewaters, and dumpsites were scattered at so many locations that the magnitude of the issue was not readily apparent. This changed with the passage of the Resource Conservation and Recovery Act in 1976 (and subsequently, CERCLA, FIFRA Amendments, and state laws) which caught waste generators of all sizes in its regulations. Specifically, RCRA (PL 94-580) required generators of acutely hazardous wastes to notify EPA and comply with facility standards by 1980, but it lacked practical guidance for compliance. With this requirement, pesticide manufacturers, formulators, applicators, farming organizations, and research institutions became mobilized, culminating in a series of symposia and national/regional workshops in the 1980s, many of which were published (*1-5*) (Table I).

The nature of the disposal issue spans a broad range:

- Empty containers
- Full or partially full containers
- Container rinsate
- Application equipment rinsate

[1]Current address: Center for Environmental Sciences and Engineering, University of Nevada—Reno, Reno, NV 89557

0097–6156/92/0510–0138$06.00/0

Table I. Abbreviated Chronology of Pesticide Waste Disposal Events, 1976-1991

1976	RCRA
1978	Pesticide Disposal Symposium (Restin, VA) Book "Disposal and Decontamination of Pesticides"
1980	RCRA Notification Deadline
1983	Symposium on Pesticide Waste Disposal (ACS, Washington, D.C.)
1984	Book "Treatment and Disposal of Pesticide Wastes"
1985	National Workshop on Pesticide Waste Disposal (Denver)
1986	Second National Workshop on Pesticide Waste Disposal (Denver)
1987	Regional Workshops on Pesticide Wastes
1988	Report "Managing Pesticide Wastes: Recommendations for Action"
1988	FIFRA—88
1989	NACA Container Management Goals
1991	Symposium on Status of the Disposal of Waste Agricultural Chemicals and their Containers (ACS, New York)

- Exterior washwater
- Contaminated soil and water at disposal sites

Thus, it was clear at the onset that a variety of approaches would be needed to deal with the overall problem. The spent container problem was the easiest to deal with, resulting in much progress. A 1991 meeting convened by the National Agricultural Chemical Association (NACA) provided several examples of responsible approaches, including container collection programs, recycling, and incineration. Special cooperative efforts have been made in several states to dispose of unwanted/unregistered pesticides in containers at a fraction of the costs required by hazardous waste disposal facilities (*6-8*).

Another clearly definable disposal problem exists with home use pesticide products. Storing these partially full containers in the garage or workshed may constitute a fire or toxic hazard, particularly if contacted by children. Thus, disposing of older materials, including nearly empty containers, at fairly frequent intervals is a worthwhile goal for many reasons. The best guidance for disposal is as follows:

- Use remaining product in accord with label directions.
- Rinse the empty container with water, adding the rinsate to the spray vessel.
- Discard the container in the refuse can or with other glass containers for recycling.

An alternative is to deliver the unwanted pesticide to a designated county or community collection station, such as have been established in New York (*9*) and a number of other states. Information on amnesty-collection programs of this type may usually be obtained from the local Cooperative Extension office, the county Health Department, or the county waste disposal facility. While the problem with homeowner disposal of pesticides is not solved, options are becoming available to the hazardous (and increasingly illegal) pouring of unused pesticides down the drain or on the soil, or loading the liquid wastes in the refuse can.

The wastewater issue also proved amenable to substantial improvement, by recycling aqueous rinsate to the application equipment tank to serve as diluent for the next application, by spraying rinsate on the crop on which the chemical is registered, or by reducing the tank residual needing rinsing through equipment modifications. Thus, early estimates of 1,000–10,000 gallons of wastewater generated each day by commercial applicators were reduced markedly. In general, once the farming community became more aware of the problem, there was rapid and positive movement to minimize the volume and scope of wastewater needing disposal.

Past heavily contaminated soils and water coupled with poorly managed application operations have compounded the problem of waste site decontamination and made it more difficult to deal with. Progress has been made with decontaminating these sites *in situ* by accelerating natural physical and biological processes (*10-11*), but dig-and-haul technology is still the mainstay of cleanup activities.

What Are Equipment Rinsewaters and Wastewater?

The nature of equipment rinsewater has been dealt with in just a few reports (*12-14*). In general, equipment tanks, hoses, pumps, booms, nozzles and exterior surfaces will contain a residue of the sprayed material which needs to be removed when changing from one chemical agent to another. This is critical if the change is from an herbicide to an insecticide or fungicide, or from a chemical which is not registered for use on the crop to be sprayed. Equipment is thus flushed internally with water, and hosed down externally, usually on a concrete pad which has drainage capability. The water generated consists of a dilute solution of the pesticide along with adjuvants, dirt, oil and miscellaneous products such as insect parts (from external washings). The procedure, and content, will differ considerably for liquid tank mixes and granular hopper contents. Considering that a typical tank mix concentration is on the order of 10,000 ppm of active ingredient, that 1-8 gallons remain as residual, and that 10-80 gallons of water are used to flush the residue, it follows that a typical first internal rinsate will contain 100-1000 ppm or 0.07-0.7 kg of active ingredient. Actual measurement of aircraft spraying a number of pesticides showed typical values (*14*) (Table II).

Table II: Typical Aircraft Wastewater Contents in ppm (*14*)

Sample	*Pydrin*	*Lorsban*	*Comite*
Tank Mix	3514	16000	15385
First Tank Rinse	112	151	1195
Second Tank Rinse	3.1	9.5	89
Third Tank Rinse	1.5	2.9	4
Exterior Wash	0.02	0.014	0.7

These results showed that increasing the number of rinsings decreased the residue load, and that the reduction was predictable given an initial mix concentration for a given equipment's tank geometry. This study was done so that aerial applicators could spray off the most heavily contaminated rinsings (at least, the first and second rinsings) so that their wastewaters needing treatment and disposal were greatly reduced.

Newer devices, such as those which provide direct injection of technical material to the spray nozzle, clearly help significantly because they eliminate the mix tank—the primary source of contaminate rinsate in the past.

Available Options for Wastewater

While there are ways to minimize pesticide contaminated water in need of disposal, it will probably never be totally eliminated—thus providing the opportunity for generating technologies which clean up or dispose of the wastewater in a legal and safe manner. How to do this best has generated a plethora of approaches, some old and a few new.

Evaporation-based approaches follow the logic of 'volume reduction' by simply allowing the water to evaporate and the increasingly concentrated residue to be attacked by sunlight, soil microbes, natural chemical forces, etc. Thus, a typical operation might have a rinse pad, a storage tank for the wastewater, and either a pond or soil containment to which the water was periodically added. Variables include:

- A sedimentation or filtration step to remove suspended materials.
- Introduction of chemical reagents or microbial amendments to the pond or soil.
- Design of the pond or soil zone, extending to the use of a leach field rather than a discrete containment.
- A removable cover over the containment to prevent intrusion of rainwater.

Hodapp and Winterlin (*10*) described a modern version of a soil bed for receiving wastewaters, and the use of amendments for accelerating decomposition. Earlier models have been described by Hall (*15*) and Winterlin (*16*). Soil beds offer the advantage of simplicity, containment, minimal loss to the air, and a matrix for slow mineralization. The disadvantages are that permit requirements may mandate double liners and monitoring wells, and a possible 'day of reckoning' will need to be faced when the containment area is cleaned out, repaired, replaced, or removed. Furthermore, evaporation pits vary in effectiveness, tending to be most efficient in areas of the U.S. with generally high water evaporation rates.

The addition of chemicals (base, oxidants, etc), energy (UV irradiation, heat), or nutrients to accelerate chemical or biological decomposition in this basic approach has provided potentially useful technologies. For example, Blankinship showed rapid degradation of MCPA-DMA in wastewater to which was added household bleach, sodium hypochlorite (*17*). Solutions of 1000 or more ppm were reduced to less than 100 ppm in 3-4 hours under outdoor sunlight conditions. Sodium perborate, a non-chlorine industrial bleach, proved useful for accelerating the decomposition of organophosphate insecticides in alkaline water. This results from the generation of HO_2^-—a 'supernucleophile' toward OPs and potentially other reactive centers (*18*).

Most OPs tested were degraded to 1/10th or less of their initial concentration in 2-3 hours of treatment. Base hydrolysis, either in bulk solution or effected by reactive columns (*19*) can also accelerate degradation of some chemical classes. Adding simple reagents to wastewater is attractive, because the pesticide is degraded and toxicity is generally reduced. However, the degradation products may prove to be less reactive and even a cause of regulatory concern themselves. Because of the differences in degradation paths for the many pesticide chemicals which exist, chemical degradation techniques can probably only be considered for well-defined wastewater.

Filtration-based treatment technologies are also attractive, perhaps because they can use off-the-shelf items such as swimming pool or industrial filters, and a variety of filtering media including ones such as peat or charcoal which sorb a variety of chemicals and are reasonably well understood and available. Charcoal-based systems have been installed at commercial applicator sites, and tested and used for several years (*20, 21*). Charcoal systems are also available for portable use to service military bases and structural pest control operators (*22*). With filtration systems, however, the key question is how to dispose of the filtered or sorbed pesticide, a question leading to at least four approaches:

- Composting (*23, 24*)
- Microbial Treatment (*25*)
- Incineration (*26*)
- Encasement in Concrete (*27*)

Of these options, incineration is the surest, and perhaps composting is the easiest, at least for on-farm applications. In one composting test 1100 ppm of simazine initially in a compost pile was reduced to 100 ppm over an 8-week period following the first addition, and from 900 to about 10 ppm in a 5-week period following the second addition of pesticides (*24*).

Some form of oxidation followed by microbial finishing of wastewater has been investigated by a number of workers. Karns et al. (*28*) began this line of study with a UV ozone system applied initially to coumaphos from cattle dip tank waste water. A more recent version (*29*) employs an ozone generator which sparges the wastewater with ozone to initiate degradation followed by percolation through a silt-loam soil column. Atrazine, cyanazine, and metolachlor were degraded quite effectively, and paraquat less so, by this system. In this study, and a more recent one on incineration (*30*), an Ames assay was used to follow the detoxification treatment. Bioassays in general have not been used very extensively for evaluating treatment options in the past. Microbial processes, including the use of the white-rot fungus, will be discussed elsewhere in this volume.

Ultrox International (*31*) reported on a UV/ozone process for treating pesticide wastewaters and also solvent contaminated groundwaters. The system included a charcoal finishing column to remove chlorinated materials not degraded by the system.

Winterlin evaluated a Perox system (UV light, hydrogen peroxide) designed to generate OH radicals for degradation of pesticides in water (*32*). The system relied more on UV than H_2O_2 to effect degradation and, as with all UV-based systems, worked much better with clear than with cloudy solutions. This is a serious drawback for UV and sunlight-based energy systems which need to interact directly with the substrate.

As these examples show, there are several techniques available now for handling wastewater. All have some limitation or disadvantage which precludes a bottom-line recommendation for all wastewater situations. Filtration followed by incineration, though expensive, is a here-and-now system adaptable to virtually any size wastestream and, until something better comes along, should be considered the standard against which other technologies should be evaluated both for efficacy and economics. Furthermore, filtration can be interfaced with a variety of other techniques. The residue can be disposed of by incineration, composting, etc., while the filtrate can be further purified by addition of a chemical reagent followed by recycling or discharge of the water (*33*).

New Technologies

New ideas are the primary subject of this symposium section and thus will not be dealt with in much detail in this paper. They fall into the categories:

- Microbial reactors
- Biotechnology approaches
- Solid-phase reactors
- Infrared incineration
- Electrolysis

Each approach has advantages and disadvantages. Microbial reactions generally can effect complete mineralization, in opposition to chemical reactions which tend to involve discrete transformations to products which may still need disposal attention. Microbial processes are, however, subject to shock if the waste stream is not fairly consistent and if conditions (temperature, pH, etc.) are outside the range of the microbial consortium employed. Biotechnology offers opportunity for improvement, by expanding the substrate range for the more prolific and hardy degraders, or by producing enzymes which can be isolated and used outside of the host organism, even in an immobilized system amenable to flow-through operations. Solid phase bioreactors are attractive for this latter reason, plus they may bypass the need for adding reagents to the waste stream which exacerbate the eventual disposal problem, and furthermore they may be regenerated. This technology is not yet available at the practical level.

Infrared incineration (along with several other forms of non-flame incineration under study for various waste disposal applications) will probably be expensive in terms of capital outlay but offers a long-term mineralizing approach to disposal of many types of organic wastes. Electrolysis seems to have been largely overlooked for disposal, even though it may have a specialty niche, such as for reductive dechlorination of halogenated organics which are resistant or not amenable to other approaches.

With regard to microbial reactors, one challenge is to provide these on a scale and in a format for ready adoption by generators of different sizes. Our laboratory is in the process of examining one commercial self-contained unit which has shown promise for degradation of organic-bearing aqueous wastes from the paint and other industries. This system accepts an aqueous waste stream into a microbial digester compartment. The gases can then be filtered and vented while the liquid effluent is recycled. This is now being tested in our laboratory with a variety of pesticide types and concentrations. As an adjunct to these tests, the effluent stream will be subjected to bioassay evaluation with the Microtox system—a bacterial-based bioassay which signals the presence of toxicants by a decrease in the viability of photoluminescent bacteria (*34*). We will also be using the Ames bioassay (*35*) and fish bioassays for effluent stream evaluations.

Conclusions

There are ample opportunities for research and development in the field of pesticide waste disposal technology, both for wastewaters (the primary focus in this chapter) and also for outdated technical material (*36*), and, certainly, for contaminated soil. In all of these situations, one might fantasize on items which are particularly desirable in the research, development, and demonstration phases:

- Simple chemical and bioassay tests to evaluate effluents
- Side-by-side comparison tests of two or more competing technologies
- Clear-cut regulatory targets, so that we will know what we are aiming for.

The third item may be unattainable given the complexity of legal and liability issues, but the first two goals can be incorporated in almost any R&D program.

Literature Cited

1. *Treatment and Disposal of Pesticide Wastes;* Krueger, R.F., Seiber, J.N., Eds., ACS Symposium Series No. 259; American Chemical Society: Washington, DC, 1984.
2. *Proceedings: National Workshop on Pesticide Waste Disposal, Denver, Colorado, January 28-29, 1985;* Bridges, J., Ed.; JACA Corp.: Fort Washington, PA, 1985, EPA/600/9-85/030.
3. *Proceedings: National Workshop on Pesticide Waste Disposal, Denver, Colorado, January 27-29, 1986;* Bridges, J., Ed.; JACA Corp.: Fort Washington, PA, 1987, EPA/600/9-87/001.
4. Gilding, T.J. *Managing Pesticide Wastes: Recommendations for Action. Summary of National Conferences and Workshops on Pesticide Waste Disposal;* National Agricultural Chemicals Association: Washington, DC, 1988.
5. *Pesticide Waste Disposal Technology;* Bridges, J.S.; Dempsey, C.R., Eds.; Pollution Technology Review No. 148; Noyes Data Corporation: Park Ridge, NJ, 1988.
6. Anon. "Bargain Basement Cleanup," *Agrichemical Age*, **December 1989**, 14-15.
7. Anon. "Pilot Disposal Project Looks Successful," *Farm Chemicals*, **April 1987**, 48-49.
8. Kozak, R.M.; Hass, E. "Difficulties Confronting an Agricultural Pesticide Waste Collection Program in Wisconsin," *Journal of Soil and Water Conservation*, **March-April 1990**, 271-272.
9. Hughes-Davis, B.A.; "Pesticide clean-up days," *The Conservationist* (State of New York, Department of Environmental Conservation), **July-August 1984**.

10. Hodapp, D.M.; Winterlin, W.L. "Pesticide Degradation in Model Soil Evaporation Beds," *Bull Environ. Contam. Toxicol.* **1989**, *43*, 36-44.
11. Craigmill, A.L.; Winterlin, W.L.; Seiber, J.N. "Biological Treatment of Waste Disposal Sites," in *Proceedings: National Workshop on Pesticide Waste Disposal, Denver, Colorado, January 27-19, 1986*; Bridges, J., Ed.; JACA Corp.: Fort Washington, PA, 1987, EPA/600/9-87/001, pp 31-37.
12. Whittaker, K.F.; Nye, J.C.; Wukasech, R.F.; Squires, R.G.; York, A.C.; Kazimier, H.A. Collection and Treatment of Wastewater Generated by Pesticide Applicators. Draft Report; Purdue University: West Lafayette, IN, 1979.
13. Taylor, A.G.; Hanson, D.; Anderson, D. "Recycling Pesticide Rinsewater," In *Proceedings: National Workshop on Pesticide Waste Disposal, Denver, Colorado, January 27-29, 1986*; Bridges, J., Ed.; JACA Corp.: Fort Washington, PA, 1987, EPA/600/9-87/001, pp 67-73.
14. Woodrow, J.E., Wong, J.M.; Seiber, J.N. "Pesticide Residues in Spray Aircraft Tank Rinses and Aircraft Exterior Washes," *Bull. Environ. Contam. Toxicol.* **1989**, *42*, 22-29.
15. Hall, C.V. "Pesticide Waste Disposal in Agriculture," in *Treatment and Disposal of Pesticide Wastes;* Krueger, R.F.; Seiber, J.N., Eds.; ACS Symposium Series No. 259; American Chemical Society: Washington, DC, 1984, pp 27-36.
16. Winterlin, W.L.; Schoen, S.R.; Mourer, C.R. "Disposal of Pesticide Wastes in Lined Evaporation Beds," in *Treatment and Disposal of Pesticide Wastes;* Krueger, R.F.; Seiber, J.N., Eds.; ACS Symposium Series No. 259, American Chemical Society: Washington, DC, pp 97-116.
17. Blankinship, M.S. Masters Thesis, "The Photodegradation of Aqueous Pesticide Waste," University of California at Davis, 1989.
18. Qian, C.; Sanders, P.F.; Seiber, J.N. "Accelerated Degradation of Organophosphorus Pesticides with Sodium Perborate," *Bull. Environ. Contam. Toxicol.* **1985**, *35*, 682-688.
19. Lemley, A.T.; Zhong, W.Z.; Janaver, G.E.; Rossi, R. "Investigation of Degradation Rates of Carbamate Pesticides," in *Treatment and Disposal of Pesticide Wastes;* Krueger, R.F.; Seiber, J.N., Eds.; ACS Symposium Series No. 259, American Chemical Society: Washington, DC, pp 245-259.
20. Antommaria, P.E. Abstract: "Pesticide Rinsewater Treatment," in *Pesticide Waste Disposal Technology;* Bridges, J.S.; Dempsey, C.R., Eds.; Pollution Technology Review No. 148; Noyes Data Corporation: Park Ridge, NJ, 1988, p 141.
21. Nye, J.C.; Way, T. "Carbon Adsorption Treatment of Rinsewater," in *Proceedings: National Workshop on Pesticide Waste Disposal, Denver, Colorado, January 27-29, 1986;* Bridges,, J., Ed.; JACA Crop.: Fort Washington, PA, 1987, EPA/600/9-87/001, pp 23-27.
22. Dennis, W.H., Jr. "A Practical System to Treat Pesticide-Laden Wastewater," in *Proceedings: National Workshop on Pesticide Waste Disposal, Denver, Colorado, January 28-29, 1985;* Bridges, J., Ed.; JACA Corp.: Fort Washington, PA, 1985, EPA/600/9-85/030, pp 49-53.
23. Mullins, D.E.; Young, R.W.; Hetzel, G.H. Abstract: "Disposal of Dilute and Concentrated Agricultural Pesticide Formulations Using Organic Matrix

Absorption and Microbial Degradation," in *Pesticide Waste Disposal Technology;* Bridges, J.S.; Dempsey, C.R., Eds.; Pollution Technology Review No. 148; Noyes, Data Corporation: Park Ridge, NJ, 1988, p 155.

24. California Agricultural Research, Inc.; *Composting for Treatment of Pesticide Rinsates;* prepared for State of California Department of Health Services Toxic Substances Control Division and US Environmental Protection Agency, 1988.
25. Craigmill, A.L.; Winterlin, W.L. "Pesticide Wastewater Disposal: Biological Methods," in *Pesticide Waste Disposal Technology;* Bridges, J.S.; Dempsey, C.R., Eds.; Pollution Technology Review No. 148; Noyes Data Corporation: Park Ridge, NJ, 1988, pp 55-72.
26. Oberacker, D.A. "Incineration Options for Disposal of Waste Pesticides, in *Proceedings: National Workshop on Pesticide Waste Disposal, Denver, Colorado, January 28-29, 1985;* Bridges, J., Ed.; JACA Corp.: Fort Washington, PA, 1985, EPA/600/9-85/030, pp 87-94.
27. Nye, J.C. "Physical Treatment Options (Removal of Chemicals from Wastewater by Adsorption, Filtration, and/or Coagulation)," in *Proceedings: National Workshop on Pesticide Waste Disposal, Denver, Colorado, January 28-29, 1985;* Bridges, J., Ed.; JACA Corp.: Fort Washington, PA, 1985, EPA/600/9-85/030, pp 43-48.
28. Karns, J.S.; Muldoon, J.T.; Kearney, P.C. "A Biological/Physical Process for the Elimination of Cattle-Dip Pesticide Wastes," in *Proceedings: National Workshop on Pesticide Waste Disposal, Denver, Colorado, January 27-29, 1986;* Bridges, J., Ed.; JACA Corp.: Fort Washington, PA, 1987, EPA/600/9-87/001, pp 43-48.
29. Somich, C.J.; Muldoon, M.T.; Kearney, P.C. "On-Site Treatment of Pesticide Waste and Rinsate Using Ozone and Biologically Active Soil," *Environ. Sci. Technol.* **1990**, *24*, 745-749.
30. DeMarini, D.M.; Houk, V.S.; Lewtas, J. "Measurement of Mutagenic Emissions from the Incineration of the Pesticide Dinoseb during Application of Combustion Modifications," *Environ. Sci. Technol.* **1991**, *25*, 910-913.
31. Ultrox International; *UV/Ozone Treatment of Pesticide and Groundwater;* prepared for State of California Department of Health Services Toxic Substances Control Division, Alternative Technology Section, 1988.
32. Winterlin, W.L.; *UV/Hydrogen Peroxide Treatment for Destruction of Pesticide Laden Waste;* prepared for California Department of Health Services Toxic Substances Control Project and US Environmental Protection Agency, 1987.
33. Seiber, J.N. "Pesticide Waste Disposal: Review and 1985 National Workshop Summary," in *Proceedings: National Workshop on Pesticide Waste Disposal, Denver, Colorado, January 27-29, 1986;* Bridges, J., Ed.; JACA Corp.: Fort Washington, PA, 1987, EPA/600/9-87/001, pp 11-19.
34. Bulick, A.A.; Greene, M.W.; Isenberg, D.L. "Reliability of the bacterial luminescence assay for determination of the toxicity of pure compounds and complex effluents," in *Aquatic Toxicology and Hazard Assessment: 4th Conference;* Branson, D.R.; Dickson, K.L., Eds.; ASTM.STP, pp 339-347.

35. Wong, J.M.; Kado, N.Y.; Kuzmicky, P.A.; Ning, H.-S.; Woodrow, J.E.; Hsieh, D.P.H.; Seiber, J.N. "Determination of Volatile and Semivolatile Mutagens in Air Using Solid Adsorbents and Supercritical Fluid Extraction," *Analytical Chemistry*, **1991**, *63*, 1644-1650.
36. Oberacker, D.A. "Test Burns for Banned Pesticides," *Journal of Hazardous Materials*, **1989**, *22*, 135-142.

RECEIVED June 8, 1992

Chapter 12

Biotechnology in Bioremediation of Pesticide-Contaminated Sites

Jeffrey S. Karns

Pesticide Degradation Laboratory, Natural Resources Institute, Agricultural Research Service, U.S. Department of Agriculture, Beltsville, MD 20705–2350

Biotechnology has been highly touted as a potential source of safe, inexpensive, and effective methods for the remediation of sites heavily contaminated with agrochemicals and for the direct treatment of agrochemical wastes. Although there have been some notable successes in the use of microorganisms for the degradation of waste chemicals, biotechnology has not yet provided a panacea for farmers or applicators. In many cases waste sites contain mixtures of chemicals, some of which may interfere with the metabolism of others. In some instances these difficulties can be overcome through selection of microorganisms or microbial consortia adapted to survival in the unique mix of chemicals present at each site. Genetic engineering may provide a means for producing microbes with the best mix of biochemical pathways for bioremediation at waste sites. Perhaps the best hope for the near-term application of biotechnology to the disposal of agrochemical wastes lies in the treatment of rinsates, equipment, and containers, where the waste is contained and its chemical composition is known.

Biotechnology, the use of biological systems for the benefit of mankind, has been around for a long time. It has been used for millennia in the food industry to make breads, brew wines and beers, and to preserve foods. Recently, the term biotechnology seems to have become synonymous with the use of the procedures commonly associated with the *in vitro* manipulation of DNA (recombinant DNA technology) for the production of products of benefit to man. Since the inception of this phase of biotechnology in the mid-1970s we have seen its application to the creation of exciting new pharmaceutical products and the development of promising new pest control products. One area in which much was envisioned of biotechnology but little has yet been realized is the remediation of chemically contaminated sites.

Biotechnology was thought to hold (and still does hold!) great potential for the development of simple and inexpensive processes for the elimination of waste chemicals. In this chapter research will be described which has shown that microorganisms are capable of degrading agricultural chemicals and work will be reviewed showing the use of microbes or microbial systems in waste disposal or remediation of contaminated sites. I will then give my impressions of why biotechnology has yet to fulfill all the promise it holds in this area and describe research that addresses some of the problems that have been experienced.

Microbial Degradation of Pesticides

There have been many reports of biological degradation of various pesticide or pesticide-like compounds. Most of the major classes of compounds currently used as pesticides are known to be subject to some form of biological transformation, and in some cases, complete mineralization. In addition, there are numerous reports of transformation and mineralization of older compounds such as DDT and lindane.

Specific vs. Non-Specific Biodegradation. As a first step in discussing the biodegradation of pesticides by microorganisms it is necessary to define what is meant by specific and non-specific mechanisms of pesticide biodegradation. By specific biodegradation I will mean degradation as a result of gene and enzyme systems that have evolved to directly degrade the particular pesticide or a very close structural relative. Examples of specific degradation are the hydrolysis of parathion by parathion hydrolase (*1*) or the complete mineralization of 2,4,5-trichlorophenoxyacetic acid by a pathway found in a strain of *Pseudomonas cepacia* (*2*). While an enzyme agent of such a system may have the ability to degrade several compounds within a class (such as parathion hydrolase's ability to degrade a number of structurally related *O,O*-dialkylphosphorothioates [*1*]) the range of substrates that can be degraded is relatively limited. Most examples of cometabolic transformation of pesticide molecules would fall under this definition.

I will define non-specific biodegradation as degradation as a result of the action of a secondary product of biological activity. Thus, chemical hydrolysis of a pesticide as a result of a pH change caused by the action of microorganism on its environment would be a good example of non-specific biodegradation. The action of peroxidase enzymes produced by lignin-degrading fungi and streptomycetes is the best documented example of such non-specific pesticide degradation mechanisms. In this case, the peroxidase enzyme reacts with hydrogen peroxide to form hydroxy radical (*3*) which then chemically attacks pesticide molecules which have bonds susceptible to attack by hydroxy radical. While there is undoubtedly some role played by the enzyme in bringing the pesticide molecule and hydroxy radical into close proximity, I consider this reaction to be of a more generic nature than those in which functional groups in the enzyme itself are responsible for substrate binding and bond breakage.

Degradation of Pesticides by Lignin-peroxidase-producing *Phanerochaete* sp.. The white-rot fungus, *Phanerochaete chrysosporium*, was one of the first organisms shown to degrade lignin through the action of a potent lignin peroxidase enzyme (*4*). This enzyme converts hydrogen peroxide into hydroxy radical which can attack lignin, breaking it into smaller components resulting in eventual decomposition of the ligneous material (*3,5,6*). It has been shown that lignin-peroxidase producing cultures of *Phanerochaete chrysosporium* can degrade a number of pesticide or pesticide-like compounds, including DDT and methoxychlor (*7,8*); lindane, chlordane and dieldrin (*8,9*); 2,4,5-T (*10*); and pentachlorophenol (*11*). In addition to *Phanerochaete* there are other white rot fungi and lignin-degrading actinomyces (*12*) which may produce lignin peroxidases that attack pesticides (*13*). Through the systematic isolation of such organisms and the characterization of the pesticide degrading ability of the lignin peroxidases they produce, it may be possible to assemble an arsenal of enzymatic weapons for use in remediation of pesticide contaminated soils.

Degradation of Pesticides by Specific Enzymes and Pathways. A large number of microorganisms which produce specific pesticide degrading enzymes or that carry entire pathways for the mineralization of pesticides have been characterized. Microbial degradation of organophosphates (*14,15*), *N*-methylcarbamates (*16,17*), triazines (*18,19*), substituted ureas (*20,21*), carbamothioates (*22,23*), phenylcarbamates (*24*), chloroacetanilides (*25,26*), glyphosate (*27*), and phenoxyacetates (*28,2*) have all been reported. Of particular interest are the broad spectrum hydrolases such as parathion hydrolase (*1,29*) and *N*-methylcarbamate hydrolases (*30,31,17*) which degrade a number of compounds within the pesticide class. Parathion hydrolase can degrade coumaphos, methyl parathion and a number of other related compounds (*29*) while the reported *N*-methylcarbamate hydrolases can degrade aldicarb, carbofuran and carbaryl and a number of other related compounds. These enzymes are stable and the genes have been cloned so that their production for use in detoxification of pesticides is feasible. While these enzymes themselves do not mineralize the pesticides they attack, the hydrolysis they catalyze does result in complete elimination of the biological activity of the compound.

Use of Biological Systems in Waste Treatment and Remediation

There have been several examples of the use of microbes for the treatment of pesticide wastes and for the remediation of contaminated sites. These treatment methods ranged from the use of indigenous microorganisms (sometimes with nutrient amendments designed to stimulate resident microbes) to the use of pure cultures of pesticide degrading organisms for the elimination of specific pesticides from wastes or from contaminated sites. Winterlin, et al. (*32*) attempted to stimulate the degradation of pesticides in soils from a site heavily contaminated with a large number of compounds. They determined the half-life of numerous pesticides under aerobic (moist) and anaerobic (saturated) conditions with and without the addition of nutrient amendments. While they did see limited degradation of some of the contaminants under certain conditions, it was obvious that it would take a long time to clean this site using such an *in situ* process. Shelton and

Hapeman-Somich (*33*) demonstrated that microorganisms indigenous to a waste solution of the acaricide coumaphos could be stimulated to accomplish the degradation of the pesticide.

Studies investigating the use of pure cultures of microbes have shown that this practice may be useful in certain situations. Several studies have examined the degradation of pentachlorophenol (PCP), a commonly used wood preservative. Edgehill and Finn (*34*) and Crawford and Mohn (*35*) demonstrated that PCP could be removed from contaminated soils by inoculation with PCP degrading bacteria. However, the latter study demonstrated some of the problems that can be encountered when it was shown that PCP was readily degraded in some soils but virtually untouched in others even though all soils received inoculation. Frick et al. (*36*) and Pflug and Burton (*37*) reported on processes based on the PCP degrading *Flavobacterium* isolated by Steiert et al. (*38*) that were useful in cleaning PCP laden groundwater. The process worked despite the presence of contamination due to creosote and other wood preservatives contaminating soils and groundwater at such locations. We have used a parathion hydrolase-producing *Flavobacterium* as part of a two step process for the elimination of coumaphos in cattle dip wastes (*39,40*). In an early field trial on 2470 L of spent cattle dip containing 1 g/L coumaphos *Flavobacterium* sp. ATCC 27551 was grown in the material after adding xylose, ammonium sulfate and buffers to create conditions suitable for growth. As shown in Figure 1 the bacterium completely hydrolyzed the coumaphos within 48 hours with concomitant accumulation of the chlorferon hydrolysis product. The chlorferon was then degraded by oxidation with ozone. More recently we have shown that the parathion hydrolase enzyme itself can be used to treat waste cattle-dip (Shelton, Hapeman-Somich, and Karns, this volume).

Role of Biotechnology in Future Waste Treatment and Remediation

Whether as *in situ* processes or as part of an engineered solution biotechnology can play an important role in the disposal of waste agrochemicals or the remediation of contaminated sites. The reasons for the failure of biotechnology to fulfill quickly all the expectations that were held are many and varied. There are few "deep pockets" associated with waste agrochemicals. The end users who end up responsible for the sites to be cleaned are usually small to medium sized businesses that cannot afford to pay for the research required to develop new technologies for the treatment of their sites. In addition, there has not been an overwhelming amount of financial support from federal or state governments. While the USDA and USEPA do fund some research in the area of pesticide degradation the amount of money spent does is not enough to support a research community large enough to gather the information needed to be able to degrade the hundreds (perhaps thousands?) of compounds that have been used as pesticides, and their environmental conversion products.

The large number of compounds that have been used as pesticides over the last 40 years presents another problem for biotechnology, that of overlapping toxicities. The site described by Winterlin, et al. (*32*) was originally characterized as containing atrazine (up to 5000 ppm), chlorpyrifos (up to 3000 ppm), diuron (up to 3,900 ppm), parathion (up to 1,900 ppm),

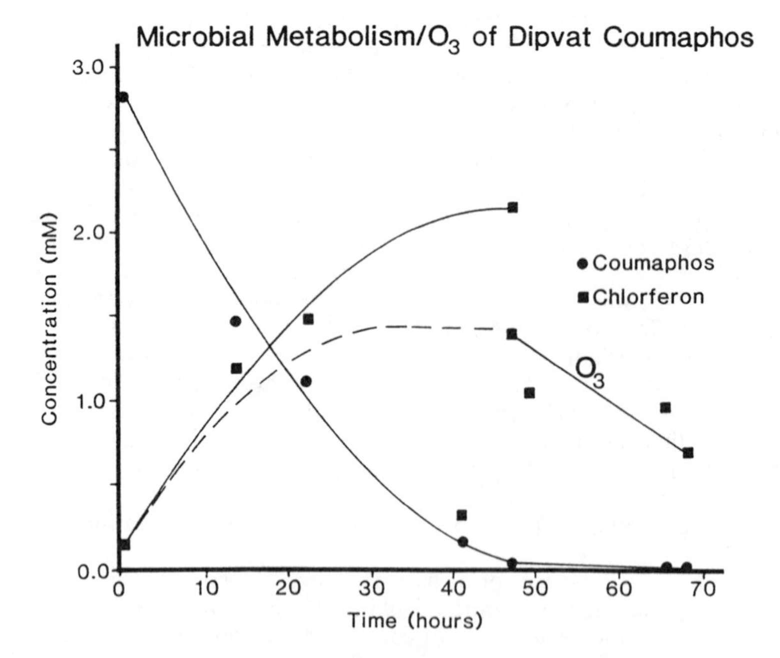

Figure 1. Degradation of the acaricide coumaphos by combined treatment with a parathion hydrolase producing *Flavobacterium* followed by oxidation of the hydrolysis product (chlorferon) with ozone. These data were obtained during a field trial in Laredo, TX and are reprinted from reference 40.

trifluralin (up to 1,100 ppm) as well as smaller amounts of 2,4-D, diazinon, methyl parathion, molinate, terbacil, thiobencarb and pronamide. When samples were taken from deeper depths older compounds such as DDT and toxaphene were found at high concentrations. This covers a broad range of compound classes all at high concentration and thus, a wide range of biochemical pathways would be required to completely degrade all of these chemicals. At these concentrations some of these chemicals might be directly toxic to microorganisms; it is also likely that some of these chemicals or intermediates in their degradation might act as inhibitors of the biochemical pathways involved in the degradation of other chemicals.

The principle of cross inhibition, and the types of things that microbiologists can do to counteract it, is illustrated in a paper by Haugland, et al. (*41*). 2,4-D and 2,4,5-T have very similar chemical structures, differing only by the addition of 1 extra chlorine atom on the aromatic ring, yet the biochemical pathways by which these compounds are known to be degraded are very different. The initial stages of these pathways are shown in Figure 2. In *Alcaligenes eutrophus* JMP134 2,4-D (compound I) is converted to 2,4-dichlorophenol (compound II) which is then converted to 3,5-dichlorocatechol (compound III). This dichlorocatechol is cleaved between the hydroxy groups (1,2- or ortho- cleavage) to yield aliphatic products which are further degraded, releasing chloride, CO_2 and H_2O (*28*). In *Pseudomonas cepacia* AC1100 2,4,5-T (compound A) is converted to 2,4,5-trichlorophenol (compound B) in a manner analogous to the first step in 2,4-D degradation, however, subsequent degradation of the trichlorophenol is through dehalogenation of the aromatic ring at the 4 and 5 positions to yield 5-chloro-1,2,4-trihydroxybenzene (compound D) which is then further metabolized through a mechanism that is presently unknown (*42*). When either organism is placed alone into a broth containing a mixture of 2,4-D and 2,4,5-T all metabolism ceases and neither compound is degraded. This is due to the fact that 2,4,5-T directly inhibits the first step in 2,4-D degradation while 2,4-dichlorophenol is mistakenly converted to 2-chloro-1,4-dihydroxybenzene (Compound IV) by the 2,4,5-T degradation pathway and this compound acts to inhibit the normal 2,4,5-T metabolism. When both organisms were inoculated together into a medium containing the mixed herbicides, the metabolism of both compounds was still inhibited. However, Haugland, et al. (*41*) showed that by moving the plasmid carrying the 2,4-D degradation pathway genes from JMP134 into AC1100 they had constructed a derivative of AC1100 that could degrade both 2,4-D and 2,4,5-T without any inhibition.

The preceding story provides an example of what can be done to overcome some of the hurdles facing the application of biotechnology to bioremediation of contaminated soils where mixtures of chemicals are likely to be the rule rather than the exception. Admittedly, there is a tremendous amount of research into biochemical pathways and genetics that must be done to be able to mix genes in the manner described to solve the problem. Such research is expensive and can take a fair amount of time. In the near term, the types of applications we are likely to see for pure cultures, enzymes, and cloned genes is in the treatment of containers, rinsates, and groundwater where the mixtures of pesticides are likely to be less complex. The most likely candidate for remediation of soils heavily contaminated with

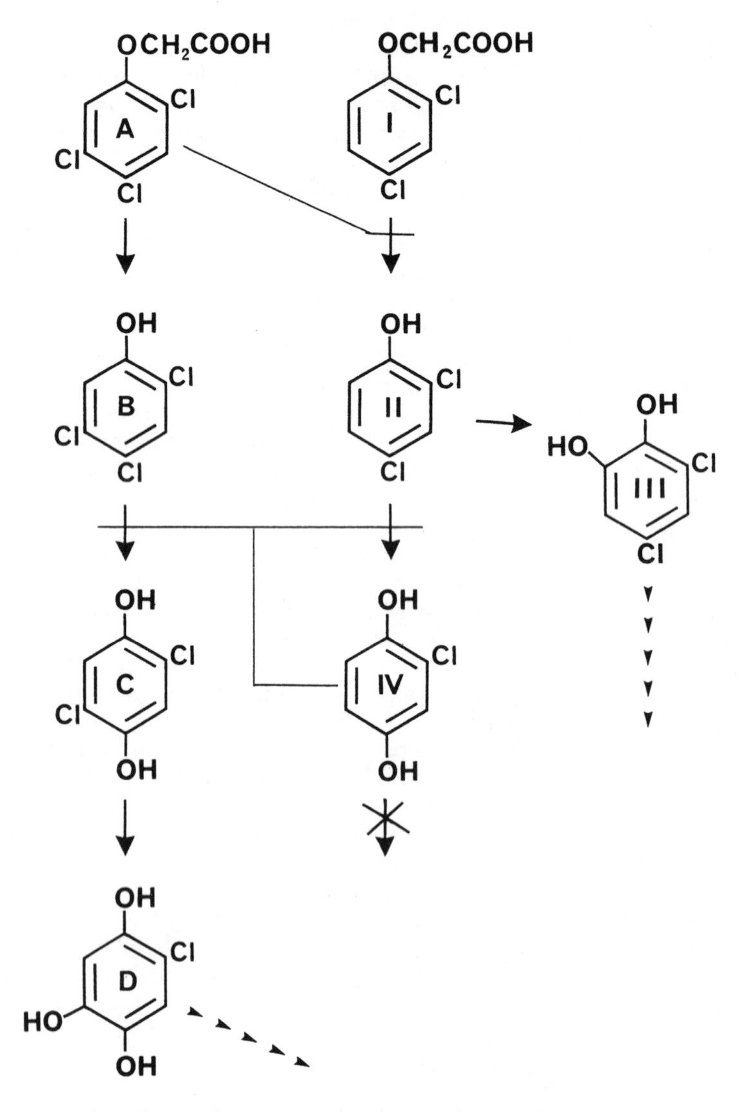

Figure 2. Pathways of metabolism of chlorphenoxyacetic acids in two distinct bacterial isolates. The pathway on the left is the 2,4,5-T degradation pathway present in *Pseudomonas cepacia* AC1100 while that on the right is the 2,4-D degradation pathway in *Alcaligenes eutrophus* JMP134. Fine lines through pathway arrows represent inhibition of that pathway step by the compound connected to the line. The X through a pathway arrow indicates that no further degradation of the metabolite is possible. This figure is adapted from data presented in reference 41.

a wide variety of pesticides is oxidative treatment with lignin peroxidase-producing organisms such as the white rot fungi or streptomycetes. The composting of contaminated soils mixed with sawdust, cornstalks, or some other ligneous material might be a viable method for removal of pesticide residues. The degradation in such systems is likely to be incomplete so that effort will have to be made to assure that the products generated are further degraded or are harmless.

In conclusion, I think that biotechnology will play an important role in the elimination of past and future agrochemical wastes. Given the remarkable ability of the microbial community to evolve the ability to degrade organic chemicals I think it just a matter of time and effort before the proper organisms are found, characterized, and utilized for waste treatment. I think that any solutions that are likely to be devised are going to combine microbiology, chemistry, and engineering and that "salt and pepper technology" (sprinkle a pack of bugs on the ground and your problems are over) is going to be the exception rather than the rule.

Literature Cited

1. Munnecke, D.M. *Appl. Environ. Microbiol.* **1976**. *32*, 7-13.
2. Kilbane, J.J.; Chatterjee, D.K.; Karns, J.S.; Kellog, S.T.; Chakrabarty, A.M. *Appl. Environ. Microbiol.* **1982**. *44*, 72-78.
3. Forney, L.J.; Reddy, C.A.; Tien, M.; Aust, S.D. *J. Biol. Chem.* **1982**. *257*, 11455-11462.
4. Kirk, T.K.; Schultz, E.; Connors, W.J.; Lorenz, L.F.; Ziekus, J.G. *Arch. Microbiol.* **1978**. *117*, 277-285.
5. Gold, M.H.; Kuwahara, M.; Chiu, A.A.; Glenn, J.K. *Arch. Biochem. Biophys.* **1984**. *234*, 353-362.
6. Tien, M.; Kirk, T.K. *Proc. Natl. Acad. Sci. USA* **1984**. *81*, 2280-2284.
7. Bumpus, J.A., Tien,M.; Wright, D.; Aust, S.D. *Science* **1985**. *228*, 1434-1436.
8. Bumpus, J.A.; Aust, S.D. *Appl. Environ. Microbiol.* **1987**. *53*, 2001-2008.
9. Kennedy, D.W.; Aust, S.D.; Bumpus, J.A. *Appl. Environ. Microbiol.* **1990**. *56*, 2347-2353.
10. Ryan, T.P.; Bumpus, J.A. *Appl. Microbiol. Biotech.* **1989**. *31*, 302-307.
11. Mileski, G.J.; Bumpus, J.A.; Jurek, M.A., Aust, S.D. *Appl. Environ.* Microbiol. **1988**. 2885-2889.
12. Adhi, T.P.; Korus, R.A.; Crawford, D.L. *Appl. Environ. Microbiol.* **1989**. *55*, 1165-1168.
13. Speedie, M.K.; Pogell, B.M.; MacDonald, M.J.; Kline, R. Jr.; Huang, Y.-I. *The Actinomycetes* **1987-1988**. *20*, 315-335.
14. Sethunathan, N.; Yoshida, T. *Can. J. Microbiol.* **1973**. *19*, 873-875.
15. Racke, K.D.; Coats, J.R. *J. Agric. Food Chem.* **1987**. *35*, 94-89.
16. Karns, J.S.; Mulbry, W.W; Nelson, J.O.; Kearney, P.C. *Pestic. Biochem. Physiol.* **1986**. *25*, 211-217.
17. Chaudhry, G.R.; Ali, A.N. *Appl. Environ. Microbiol.* **1988**. *54*, 1414-1419.
18. Cook, A.M. *FEMS MIcrobiol. Rev.* **1987**. *46*, 93-116.
19. Giardina, M.C.; Giardi, M.T.; Filacchione, G. *Agric. Biol. Chem.* **1982**. *46*, 1439-1445.

20. Englehardt, G.; Wallnofer, P.R.; Plapp, R. *Appl. Microbiol.* **1973**. *26*, 709-718.
21. Joshi, M.M.; Brown, H.M.; Romesser, J.A. *Weed Sci.* **1985**. *33*, 888-893.
22. Mueller, J.G.; Skipper, H.D.; Kline, E.L. *Pestic. Biochem. Physiol.* **1988**. *32*, 189-196.
23. Tam, A.C.; Behki, R.M.; Kahn, S.U. *Appl. Environ. Microbiol.* **1987**. *53*, 1088-1093.
24. Kaufman, D.D. *J. Agric. Food Chem.* **1967**. *15*, 582-591.
25. Tiedje, J.M.; Hagedorn, M.L. *J. Agric. Food Chem.* **1975**. *23*, 77-82.
26. Villarreal, D.T.; Turco, R.F.; Konopka, A. *Appl. Environ. Microbiol.* **1991**. *57*, 2135-2140.
27. Shinabarger, D. L.; Braymer, H.D. *J. Bacteriol.* **1986**. *168*, 702-707.
28. Don, R.H.; Weightman, A.J.; Knackmuss, H.-J.; Timmis, K.N. *J. Bacteriol.* **1985**. *161*, 85-90.
29. Brown, K.A. *Soil Biol. Biochem.* **1980**. *12*, 105-112.
30. Karns, J.S.; Tomasek, P.T. *J. Agric. Food Chem.* **1991**. *39*, 1004-1008.
31. Mulbry, W.W; Eaton, R.W. *Appl. Environ. Microbiol.* **1991**. *57*, 3679-3682.
32. Winterlin, W.; Seiber, J.N.; Craigmill, A.; Baier, T.; Woodrow, J.; Walker, G. *Arch. Environ. Contam. Toxicol.* **1989**. *18*, 734-747.
33. Shelton, D.R.; Hapeman-Somich, C.J. In *On-Site Bioreclamation;* Hinchee, R.E.; Olfenbuttle, R.F., Ed; Butterworth-Heinemann: Stoneham, MA, **1991**, pp 313-323.
34. Edgehill, R.U.; Finn, R.K. *Appl. Environ. Microbiol.* **1983**. *45*, 1122-1125.
35. Crawford, R.L.; Mohn, W. *Enzyme Microb. Technol.* **1985**. *7*, 617-620.
36. Frick, T.D.; Crawford, R.L.; Martinson, M.; Chresand, T.; Bateson, G. In *Environmental Biotechnology;* Omenn, G.S., Ed.; Plenum Press, New York and London, **1987**, pp 173-191.
37. Pflug, A.D.; Burton, M.B. In *Environmental Biotechnology*; Omenn, G., Ed.; Plenum Press, New York and London, **1987**. pp 193-201.
38. Steiert, J.G.; Crawford, R.L. *Biochem. Biophys. Res. Commun.* **1986**. *141*, 825-830.
39. Kearney, P.C.; Karns, J.S.; Muldoon, M.T.; Ruth, J.M. *J. Agric. Food Chem.* **1986**. *34*, 702-706.
40. Karns, J.S.; Muldoon, M.T.; Mulbry, W.W.; Derbyshire, M.K.; Kearney, P.C. In *Biotechnology in Agricultural Chemistry;* LeBaron, H.M.; Mumma, R.O.; Honeycutt, R.C.; Duesing, J.H., Eds.; ACS Symposium Series *334*, Amer. Chem. Soc., Washington, DC, **1987**, pp 156-170.
41. Haugland, R.A.; Schlemm, D.J.; Lyons, R.P.; Sferra, P.R.; Chakrabarty, A.M. *Appl. Environ. Microbiol.* **1990**. *56*, 1357-1362.
42. Sangodkar, U.M.X.; Aldrich, T.L.; Haugland, R.A.; Johnson, J.; Rothmel, R.K.; Chapman, P.J.; Chakrabarty, A.M. *Acta Biotechnol.* **1989**. *9*, 301-316.

RECEIVED April 21, 1992

Chapter 13

Chemical Degradation of Pesticide Wastes

Cathleen J. Hapeman-Somich

Pesticide Degradation Laboratory, Natural Resources Institute, Agricultural Research Service, U.S. Department of Agriculture, Beltsville, MD 20705

The principal degradation pathways for pesticides involve photolysis, hydrolysis, dehalogenation and oxidation. Although many of these reactions have been examined extensively in the literature using technical grade material, all too often transposition from the laboratory to the field has not been straightforward. Formulating agents can act as buffers or inhibitors in hydrolysis or dehalogenation. In photolysis, a light source inappropriate for field use may have been used, the formulating agents may absorb the photon energy or the solution may have been too opaque for light penetration. Oxidation processes are not always effective for all compounds, particularly organochlorines. In addition, formulating agents and surfactants can quench or remove the oxidative species. Recent field studies have also shown that fertilizers present in pesticide rinsates can deter degradation. In the past, disappearance of parent material was considered to be evidence for complete degradation, but clearly, products are formed whose toxicities and overall fate must be considered.

Elimination of pesticide wastes has received considerable attention over the past several decades as evidenced by an increase in research and the onset of governmental regulations. The agricultural community has become increasingly aware of the environmental impact of pesticides and the problems associated with improper disposal. The merits of depositing excess pesticide and equipment rinsate on the soil to degrade has become a liability. More recently simple detoxification of parent materials, which is generally a one-step, rapid process, has also come under scrutiny because the products of such techniques may be unknown, poorly understood or toxic themselves.

The optimum goal of any disposal technique should be elimination of all environmental toxicity. The disposal strategy, often multifaceted, should provide complete mineralization of the pesticide such that the only products remaining are carbon dioxide, salts, water, phosphates, nitrates, etc. While this may seem to be a severe requirement, lack of adequate consideration of the environmental impact of disposal methods has put our nation's water supply in jeopardy. One must assess all risks associated with each option to determine the most appropriate approach.

Chemical treatment processes that have been considered for pesticide waste treatment are photolysis, hydrolysis, dehalogenation and oxidation. Dissipation of the parent compounds has been the main criterion used to measure the success of the method. For the most part, these techniques can detoxify waste but, until recently, final disposition of the products formed has not been adequately considered. The criteria for choosing a chemical treatment technology should include the level of disposal desired, the functional groups of the active ingredients, and the chemical requirements of the reagents. The concentration of the pesticides, what formulating agents and surfactants have been used and the matrix of pesticide, i.e., soil, water, detergents or organic solvents, are additional considerations. All of these factors can impact the rate and success of the selected process.

Photolysis

The most important condition for photolysis is that the absorption spectrum of the pesticide must overlap the emission spectrum of the light source in order for a molecular change to occur. For most pesticides this dictates the use of high energy lamps, wavelengths less than 254 nm, and precludes direct environmental photolysis since the energy of solar radiation is too low, that is, wavelengths at the earth's surface are greater than 280 nm *(1)*. Photosensitizers, such as rose bengal, methylene blue or riboflavin, can absorb lower energy light and have been used to overcome this impediment, permitting the use of longer wavelength lamps. This is similar to the sensitization processes involving humic substances in soil which can absorb solar energy *(2)*.

Photolysis of most pesticides typically affords products that are dechlorinated and/or more oxidized, via homolytic cleavage, hydroxyl substitution or an electron transfer process. Often these products are more readily degraded by indigenous soil microbes *(3-6)*. While photolytic reactions under laboratory conditions with analytical grade material may be rapid, field studies with formulated materials have yielded mixed results *(4,7)*. The micellar action of nonionic surfactants formulated with some herbicides has been shown to increase photodegradation rates and change the product ratios *(8)*. On the other hand, interference from particulate matter, surfactants and formulations can severely decrease the intensity of light able to penetrate the solution. Humic materials and some additives can competitively absorb the available light, acting as photosensitizers and increasing the pesticide decomposition rate. However, if these compounds are not efficient sensitizers, a decrease in the reaction rate will be observed.

Hydrolysis

Perhaps one of the simplest treatment methods for pesticide wastes, considered by a number of researchers, has been treatment with caustic lime *(9-12)*. Pyrethroids, carbamates, organophosphates and acetanilides can be hydrolyzed under a variety of conditions. Unfortunately, formulations used in the commercial product sometimes behave as a buffering agent and inhibit hydrolysis *(9,12)*; thus, direct extrapolation of laboratory studies to field is not always possible.

For some pesticides, the conditions of hydrolysis afford other than the desired products. Alkaline hydrolysis of malathion in water, for example, promotes cleavage of the phosphorus-sulfur bonds to give mercaptosuccinate whereas hydrolysis in organic solvents (such as occur in ULV formulations) favors β-elimination to give fumarate *(13)*. A number of hydrolysis products may be fairly susceptible to microbial mineralization, as is presumably the case for malathion, but this may not be true for other pesticides. For example, coumaphos hydrolysis produces chlorferon which has been observed to inhibit microbial activity *(14)*.

Dehalogenation. One specific application of hydrolysis is dehalogenation, which utilizes polyethylene glycol with base, sometimes at elevated temperatures *(15-17)*. This method has been examined for a number of halogenated hydrocarbons including dieldrin, lindane, 2,4-D, 2,4,5-T and EDB and has been shown to readily remove small quantities from soil. This technology yields products that are less halogenated and more oxidized along with ethyoxylated products which significantly increase the molecular mass. Recently, a variation of this process was considered by other investigators for mineralizing PCB's; however, further testing revealed that the vast majority of the material was actually volatilized *(18)*. This further demonstrates the need to fully evaluate chemical disposal techniques before applying them in actual waste situations.

Oxidation Techniques

The limiting factor in the previous processes is the specificity of the reaction. Oxidation processes, via super oxidizing species such as hydroxy radical ($OH^{\cdot}$), are much less specific, typically removing active hydrogens forming alcohols, carbonyls and acids. Reactive sites include alkyl chains adjacent to electron donating moieties and aromatic rings that are not electron deficient. Complete removal and/or oxidation of the alkyl side chains are usually observed as well as oxidation and/or cleavage of the aromatic rings *(6)*. Of all the pesticide groups, organochlorines, are not expected to be particularly reactive towards oxidation and little research has been reported. Dechlorination has occasionally been observed *(6, 19)*, but formation of these products can often be explained by secondary processes, not necessarily arising from direct reaction with the oxidant.

The pesticide active ingredient is not the only compound present in formulated products that can react with $OH^{\cdot}$. The so-called inerts, that is, formulation ingredients, synergists and surfactants, often possess long alkyl chains and are commonly the primary component of the commercial product. Thus, the concentration and the structure of these inerts may have a significant impact on the overall rate of degradation, scavenging the $OH^{\cdot}$ initially, which retards the degradation rate but at the same time generating a peroxide radical which may increase the pesticide degradation rate overall.

Generation of hydroxy radical can be achieved using hydrogen peroxide (added or produced *in situ*) with a variety of catalysts and reagents, which include iron salts, titanium dioxide, ozone, ultraviolet light, electrochemical precipitation and hydroxide (pH control). An intricate relationship exists between concentrations of $OH^{\cdot}$, H_2O_2, RH_2 (the pesticide, formulating agent or other oxidizable organic species), OH^- and various catalysts; thus the pK_a's and the rate constants of the individual reactions determine the efficiency of the overall process *(20)*.

Ozonation processes. The balance between the various reactive species, oxidizing agents and reaction conditions of aqueous ozonation has been studied extensively *(20)*. At high pH, H_2O_2 deprotonates (pK_a = 11.6) and the resultant conjugate base, peroxy anion HO_2^-, reacts with ozone to give ozonide, $^{\cdot}O3^-$, and peroxy radical, $^{\cdot}HO_2$. The pK_a of $^{\cdot}HO_2$ is 4.8 and under alkaline conditions it dissociates to $^{\cdot}O_2^-$. This radical anion rapidly reacts with ozone relative to organic species and via electron transfer also affords ozonide which decomposes to $OH^{\cdot}$ *(21, 22)*. Hydroxy radical then reacts with RH_2 forming an organic radical, $^{\cdot}RH$, which reacts with O_2 to give peroxy radical, $^{\cdot}O_2RH$ *(23, 24)*. Subsequent decomposition affords oxidized R and superoxide, $^{\cdot}O_2^-$. Alternatively, two $^{\cdot}O_2RH$ can couple to give a tetroxide intermediate, $R_2O_4H_2$, which in a poorly understood mechanism gives rise to H_2O_2 and oxidized R *(25, 26)*. Because H_2O_2 is generated, the ozone decomposition and hydroxy radical production cycle is continued; however a sudden burst of any one of the reagents will disrupt the efficiency of the process (Figure 1). The low solubility of ozone in water also limits the overall rate of reaction.

Many investigators have examined the usefulness of ozone with UV or hydrogen peroxide to degrade a multitude of pollutants *(27)* and several are now applying this technology to decontaminate pesticide containing waste streams *(6, 28)*. The effect of ozone on several organochlorines has been examined and aside from oxidation of the double bond, no other structural changes were noted. *(29)*. Organophosphates react with ozone but more toxic intermediates can be formed if the reaction is not continued beyond this stage *(30)*. Phenoxyalkyl acids and esters, acetanilides and *s*-triazines are all reactive towards ozonation *(6, 31-33)*.

Photolytic ozonation, which combines both the effects of direct photolysis and ozonation, enhances the production of $OH^{\cdot}$ since photolysis of ozone gives rise to H_2O_2. Thus, UV/O_3 is more efficient than the sum of the two processes individually *(20)*. But, because most pesticide wastes are often opaque, this technique will have limited usefulness in the field.

Titanium dioxide and iron complexes. One of the first methods of hydroxy radical decomposition of pesticides utilized Fenton's reagent (ferrous salts and hydrogen peroxide) relying on iron complexes to shuttle electrons (equations

1, 2). In these initial experiments, atrazine was dealkylated and the ring of amitrole was cleaved *(34, 35)*. The use of iron complexes with hydrogen peroxide and the influence of UV irradiation on the process has recently been

$$Fe^{2+} + H_2O_2 \dashrightarrow Fe^{3+} + {}^{\cdot}OH + OH^{-} \quad (1)$$

$$^{\cdot}OH + RH_2 \dashrightarrow {}^{\cdot}RH + H_2O \quad (2)$$

reexamined as a means of degrading atrazine *(36)*. Several other groups have initiated studies utilizing UV light in conjunction with iron salts, but as with other photolytic processes, light attenuation in field situations may decrease the reaction rate or inhibit the reaction entirely.

Titanium dioxide has also been utilized as a photocatalyst with a simulated solar light source, generating conditions that appear to resemble hydroxy radical oxidation. Under laboratory conditions several *s*-triazines were dealkylated, deaminated and dechlorinated *(19)*. This technique can potentially be effective for a broad spectrum of pesticides; however, application at this time appears to be limited to ground and surface waters that are not opaque.

Combined Chemical and Biological Treatment

Chemical transformations have been shown to change the inherent biodegradability of some xenobiotics. Aqueous photolysis of 2,4-dichlorophenol and 2,4,5-trichlorophenol in the presence of hydrogen peroxide greatly enhanced the biodegradation *(37)*. Photolysis of humic acid bound glycine increased the mineralization rate *(38)*. Pretreatment of certain organophosphates with sodium perborate accelerated biodegradation *(39)*. Ozonation of a number of pesticides has been shown to significantly enhance the rate of mineralization under laboratory conditions *(40-42)*. Field tests demonstrated the potential usefulness of a binary process whereby pesticide waste was ozonated and then passed through a bioreactor filled with highly organic soil (Figure 2). Substantial fluctuations in the types and concentrations of pesticides affected the ozonation efficiency. First order degradation rates were not observed as were found in the laboratory using technical grade material, suggesting that in field tests surfactants and adjuvants play an important role in the degradation process *(41)*. Additional studies showed that monitoring microbial activity was difficult due to interference from soil components, and that high fertilizer concentrations typically found in pesticide wastes inhibited microbial degradation *(42, 43)*. Further research has led to isolation of more resilient organisms and use of an inert bioreactor support matrix *(43)*.

Conclusion

The best strategy for pesticide waste disposal begins at the generator level. Economic and regulatory issues have given risen to pesticide management schemes that will eliminate waste in some arenas and minimize the amount generated in others. While this has decreased the demand for onsite disposal,

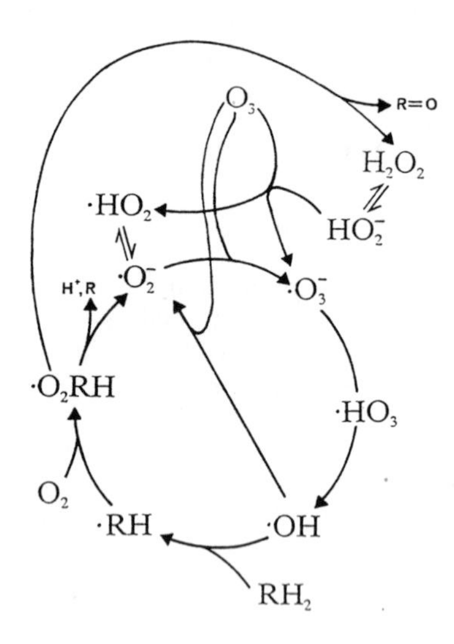

Figure 1. Decomposition of ozone to form hydroxy radical.

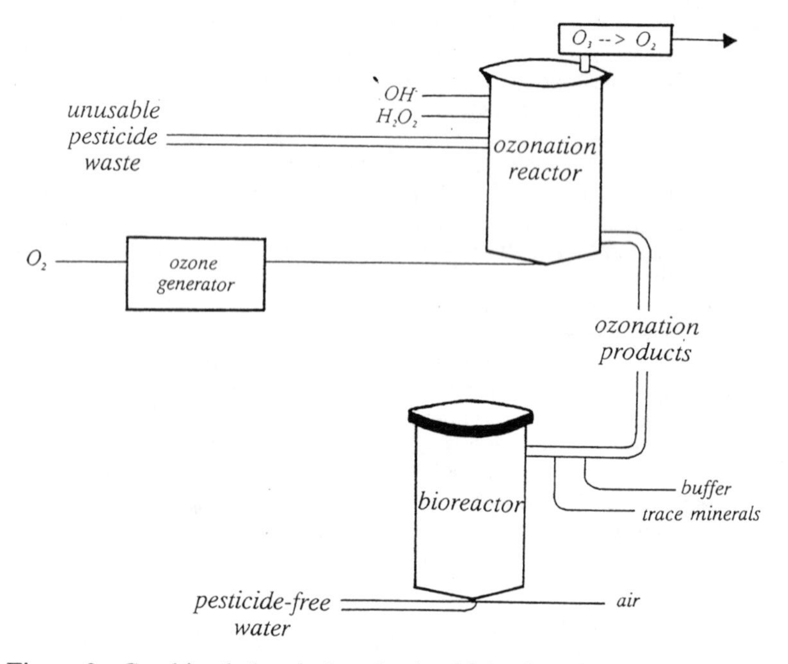

Figure 2. Combined chemical and microbial mineralization scheme to treat pesticide waste.

remediation of unusable wastes and contaminated areas is still needed. The disposal level, the components in the waste and the fate and toxicity of the products must be examined when evaluating a chemical treatment process. Hydrolysis and processes involving photolytical techniques are very useful for homogeneous systems. Alternatively, oxidation processes involving highly active species such as hydroxy radical have been shown to be more generic and can be adapted for mixtures of compounds. Such techniques coupled with either biological processes to complete mineralization, or carbon filtration to remove residual contaminants show the most promise, yet optimizing these schemes to adapt to the needs of agriculture will require more research.

Literature Cited

1. Crosby, D.G. In *Herbicides. Chemistry, Degradation and Mode of Action*; Kearney, P.C.; Kaufman, D.D., Eds.; Marcel Dekkar: New York, NY, 1976, Vol. 2; pp 825-890, and references therein.
2. Miller, G.C.; Hebert, V.R. In *Fate of Pesticides in the Environment. Proceedings of a Technical Seminar*; Biggar, J.W.; Seiber, J.N., Eds.; Publication 3320; University of California, Division of Agricultural and Natural Resources: Oakland, CA, 1987; pp 75-86.
3. Kearney. P.C.; Plimmer, J.R.; Li, Z.M. In *Pesticide Chemistry. Human Welfare and the Environment*; Miyamoto, J., Ed.; Pergamon Press: New York, NY, 1983, pp 397-400.
4. Kearney, P.C.; Zeng, Q.; Ruth, J.M. In *Treatment and Disposal of Pesticide Wastes*; Krueger, R.F.; Seiber, J.N., Eds.; ACS Symposium Series No. 259; American Chemical Society: Washington, DC, 1984; pp 195-209.
5. Somich, C.J.; Kearney, P.C.; Muldoon, M.T.; Elsasser, S. *J. Agric. Food Chem.* **1988**, *36*, 1322-1326.
6. Hapeman-Somich, C.J. In *Pesticide Transformation Products: Fate and Significance in the Environment*; Somasundaram, L.; Coats, J.R., Eds.; ACS Symposium Series No. 459; American Chemical Society: Washington, DC, 1991; pp 133-147.
7. Kearney, P.C.; Muldoon, M.T.; Somich, C.J. *Chemosphere* **1987**, *16*, 2321-2330.
8. Tanaka, F.S. In *Adjuvants and Agrochemicals*; Chow, P.N.P, Ed.; CRC Press: Boca Raton, FL, 1989, Vol. 2; pp 15-24.
9. Hemley, A.T.; Zhong, W.Z.; Janauer, G.E.; Rossi, R. In *Treatment and Disposal of Pesticide Wastes*; Krueger, R.F.; Seiber, J.N., Eds.; ACS Symposium Series No. 259; American Chemical Society: Washington, DC, 1984; pp 245-259.
10. Lande, S.S. Identification and Description of Chemical Detoxification Methods for the Safe Disposal of Selected Pesticides. NTIS PB-285208. Springfield, VA. 1978.
11. Schmidt, C.; Klubek, B.; Tweedy, J. In *Proceedings: National Workshop on Pesticide Disposal*; Bridges, J.S., Ed.; EPA/600/8-87/001. 1987. pp 45-55.
12. Shelton, D.R.; Hapeman-Somich, C.J. Pesticide Degradation Laboratory, USDA/ARS, unpublished data.

13. Eto, M. Organophosphorus Pesticides: Organic and Biological Chemistry. CRC Press: Cleveland, OH, 1974; pp 57-82, and references therein.
14. Shelton, D.R.; Somich, C.J. *Appl. Environ. Microbiol.* **1988**, *54*, 2566-2571.
15. Tiernan, T.O.; Wagel, D.J.; Vanness, G.F.; Garnett, J.H.; Solch, J.G.; Rogers, C. *Chemosphere* **1989**, *19*, 573-578.
16. Rogers, C.J.; Kornel, A.; Sparks, H.L. "Catalytic Dehydrohalogenation. Chemical Destruction of Halogenated Organics". *Abstracts of Papers*, 198th National Meeting of the American Chemical Society, Miami, Florida, September 10-15, 1989; American Chemical Society: Washington, DC, 1989, ENVR 157.
17. Rogers, C.J.; Kornel, A. "Chemical Destruction of Halogenated Aliphatic Hydrogens", **1986**, U.S. Patent No. 4675464.
18. Anonymous *Chem. Engineering News* **1991**, Nov 11, p 18 and references therein.
19. Pelizzetti, E.; Maurino, V.; Minero, C.; Carlin, V.; Pramauro, E.; Zerbinati, O.; Tosato, M.L. *Environ. Sci. Technol.* **1990**, *24*, 1559-1565.
20. Peyton, G.R.; Glaze, W.H. In *Photochemistry of Environmental Aquatic Systems*; Zika, R.G.; Cooper, W.J., Eds.; ACS Symposium Series No. 327; American Chemical Society: Washington, DC, 1987; pp 76-88.
21. Buhler, R.E.; Staehelin, J.; Hoigne, J. *J. Phys. Chem.* **1984**, *88*, 2560-2564.
22. Staehelin, J.; Buhler, R.E.; Hoigne, J. *J. Phys. Chem.* **1984**, *88*, 5999-6004.
23. Staehelin, J.; Hoigne, J. *Environ. Sci. Technol.* **1982**, *16*, 676-681.
24. Staehelin, J.; Hoigne, J. *Environ. Sci. Technol.* **1985**, *19*, 1206-1213.
25. Peyton, G.R.; Smith, M.A.; Peyton, B.M. Photolytic Ozonation for Protection and Rehabilitation of Ground Water Resources: A Mechanistic Study. Water Resources Center Research Report 206; University of Illinois: Champaign, IL, 1987.
26. Peyton, G.R.; Gee, C.S.; Smith, M.A.; Brandy, J.; Maloney, S.W. In *Biohazards of Drinking Water Treatment*; Larson, R.A., Ed.; Lewis Publishers: Chelsea, MI, 1989; pp 185-200.
27. Glaze, W.H. *Environ. Sci. Technol.* **1987**, *21*, 224-230.
28. Richard, Y.; Brener, L. In *Handbook of Ozone Technology and Applications*; Rice, R.G.; Natzer, A., Eds.; Butterworth Publishers: Boston, MA, 1984; pp 77-97.
29. Erb, F.; Dequidt, J.; Dourlens, A.; Pommery, J.; Brice, A. *Tech. Sci. Municipales* **1979**, *13*, 168-172.
30. Richard, Y.; Martin, G.; Laplanche, A. *Tech. Sci. Municipales* **1972**, *6*, 271-273.
31. Dore, M.; Langlais, B.; Legube, B. *Water Res.* **1978**, *12*, 413-425.
32. Legube, B.; Guyon, S.; Dore, M. *Ozone Sci. Engineering* **1987**, *9*, 233-246.
33. Hapeman-Somich, C.J.; Zong, G.-M.; Lusby, W.R.; Muldoon, M.T.; Waters, R. "Aqueous Ozonation of Atrazine. Product Identification and Description of Degradation Pathway". (manuscript submitted to *J. Agric. Food Chem.* 1991).
34. Plimmer, J.R.; Kearney, P.C.; Kaufman, D.D.; Guardia, F.S. *J. Agric. Food Chem.* **1967**, *15*, 996-999.
35. Plimmer, J.R.; Kearney, P.C.; Klingebiel, U.I. *J. Agric. Food Chem.* **1971**, *19*, 572-573.
36. Larson, R. *J. Agric. Food Chem.* **1991**, *39*, 972-977.

37. Miller, R.M.; Singer, G.M.; Rosen, J.D.; Bartha, R. *Environ. Sci. Technol.* **1988**, *22*, 1215-1219.
38. Amador, J.A.; Alexander, M.; Zika, R.G. *Appl. Environ. Microbiol.* **1989**, *55*, 2843-2849.
39. Qian, C.F.; Sanders, P.F.; Seiber, J.N. *Bull. Environ. Contam. Toxic.* **1985**, *35*, 682-688.
40. Karns, J.S.; Muldoon, M.T.; Mulbry, W.W.; Derbyshire, K.; Kearney, P.C. In *Biotechnology in Agricultural Chemistry*; LeBaron, H.M.; Mumma, R.O.; Honeycutt, R.C.; Duesing, J.H., Eds.; ACS Symposium Series No. 334; American Chemical Society: Washington, DC, 1987; pp 156-170.
41. Somich, C.J.; Muldoon, M.T.; Kearney, P.C. *Environ. Sci. Technol.* **1990**, *24*, 746-749.
42. Hapeman-Somich, C.J.; Shelton,D.R.; Leeson, A.; Muldoon, M.T.; Rouse, B.J. "Field Studies using Ozonation and Biomineralization as a Disposal Methodology for Pesticide Wastes". (manuscript to be submitted to *Chemosphere* 1992)
43. Leeson, A.; Hapeman-Somich, C.J.; Shelton, D. "Biomineralization of Atrazine Ozonation Products." (manuscript to be submitted to *J. Agric. Food Chem.* 1992)

RECEIVED March 30, 1992

Chapter 14

Pesticide Wastewater Cleanup Using Demulsification, Sorption, and Filtration Followed by Chemical and Biological Degradation

Donald E. Mullins[1], Roderick W. Young[2], Glen H. Hetzel[3], and Duane F. Berry[4]

[1]Department of Entomology, [2]Department of Biochemistry and Nutrition, [3]Department of Agricultural Engineering, and [4]Department of Crop and Soil Environmental Sciences, Virginia Polytechnic Institute and State University, Blacksburg, VA 24061

Contamination of soil, surface and groundwater due to pesticide usage and improper disposal is becoming increasingly problematic for applicators of hazardous pesticide materials. A variety of disposal methods have been examined, some of which are effective, but most are either too costly or involve complicated procedures or equipment to be practical. We are developing a pesticide disposal method utilizing demulsification, sorption and filtration as a means to remove pesticides from aqueous suspensions. Once removed from rinsate or runoff solutions, sorbed pesticide is placed in bioreactors where degradation occurs during solid state fermentation (ie. composting). Successful completion of this work should provide a practical, effective, safe and inexpensive method for dilute and concentrated pesticide waste disposal.

Pesticide use in U.S. agriculture has resulted in several public and environmental concerns including: 1) the fate of their residues in the environment once they have been applied, 2) the need for an appropriate disposal method for unused, concentrated and dilute pesticide formulations and pesticide-contaminated products, and, 3) the need for methods for dealing with pesticide spills. These and other related problems are of increasing concern since pesticides are finding their way into our water supplies. Environmental contamination due to pesticide usage and disposal is becoming increasingly problematical for applicators of hazardous pesticide materials (**1,9**). A shift in philosophy regarding disposal of hazardous waste is emerging. This change involves the concept of neutralizing or degrading toxic chemicals to a point where their residues or byproducts are non-toxic and will not pose a threat to the environment (air, ground water, etc.) should they become dispersed. Current methods available for reduction of pesticide waste generation include: waste minimization, recycling rinsates, on site respraying of

0097–6156/92/0510–0166$06.00/0

rinsates, volume reduction in evaporation/degradation pits and carbon sorption. A variety of disposal methods have been examined. Some of these are effective, but most are either too costly or involve complicated procedures or equipment (**1,9**). It is essential that pesticide disposal methods be effective, safe, inexpensive and relatively easy to initiate. Disposal methods which have these attributes will fall into much broader use than those which do not. One approach which shows promise utilizes biological agents (microbes) to degrade hazardous materials to non-toxic byproducts (**3**). We have been examining the feasiblity of using sorption linked with chemical and biodegradation as a means for neutralizing liquid hazardous pesticide wastes. Treatment of solutions containing initial concentrations of diazinon or chlorpyrifos ranging from 1,250 to 20,000 mg/L with peat moss resulted in reductions ranging from 55% to 99.3%. (**5**). Studies using 1 to 6 cubic foot bioreactors have demonstrated that chemical and biological degradation of pesticides (chlorpyrifos, diazinon, carbofuran, chlordane and metolachlor) sorbed onto lignocellulosic materials may prove to be effective in destroying pesticide wastes (**5,7**).

There are currently 200 pesticidal active ingredients used in a large number of formulations, 75% of which are used as liquid sprays (**16**). These water-based liquid sprays are formed from mixtures of emulsifiable concentrate, wettable powder, soluble liquid and suspension concentrate formulations (**13,16**). Emulsifiable concentrates, wettable powders and soluble liquids are currently the most commonly used formulations for liquid sprays. However, suspension concentrates, capsule emulsions, water dispersible granules, emulsions in water and suspoemulsions will likely increase in use at the expense of traditional formulations (**13**). Changes in formulation development and availability are resulting from recent reassessment of pesticide inert ingredients (**13,15**). Because of these trends, development of waste disposal strategies should be designed to accommodate formulations currently available as well as formulations resulting from the newer technologies. The pesticide disposal process under development and which is described here, is designed to remove pesticides from a variety of aqueous spray solutions and to dispose of them using chemical and/or biological degradation.

Biologically-based Disposal System Model

A biologically-based system for a pesticide wastewater disposal process has been proposed (**5**). The process includes both a sorption and a disposal phase. The sorption phase utilizes demulsification agents and lignocellulosic materials (peat moss, wood products) to remove solubilized pesticides or their suspensions from the aqueous phase onto organic sorbents. Demulsification facilitates the sorption process when treating emulsifiable concentrate formulations (**6**). After sorption and physical separation of the pesticide-laden sorbents, the disposal phase includes placement of the solids containing pesticides sorbed onto the lignocellulosic matrices into a composting environment where the pesticides are degraded. Lignocellulosic materials (peat moss, and steam-exploded wood fibers) are being used in this disposal process because they are inexpensive, can be highly sorbent to pesticide materials and can support microbial activities associated with pesticide degradation (**8**).

Microbial degradation of organic wastes (under composting or solid state fermentation conditions) is being used in municipal waste disposal (**3,11**). Its potential for disposal of hazardous wastes has also been suggested (**3,18**). Reliance upon direct metabolism or cometabolism by various composting microorganism populations functioning as a consortia may provide for effective pesticide degradation (**8**).

Methodology

All solvents used in this study were pesticide grade and analytical standards were obtained from USEPA Pesticide and Industrial Chemicals Repository MD-8 Research Triangle Park, NC. The general pesticide extraction and analytical procedures are described by Watts **(17)**. Depending upon the pesticides, several different solvents were used as extractants (ie. acetone, hexane, etc.) and sonication was employed as a means to improve extraction efficiencies. Following various cleanup and volume reduction procedures, samples were analyzed using either gas-liquid or high performance liquid chromatography. Sorbents which were tested included sphagnum peat moss and steam-exploded yellow poplar wood fibers **(10)**, both of which were ground in a Wiley mill using a 2 mm screen. Activated carbon (Calgon Filtrasorb 200) was used as a control for sorption comparisons.

One-step demulsification, sorption and filtration solution cleanup. A 100 milliliter aqueous solution containing approximately 5000 mg/L pesticide was mixed with 2 grams sorbent and 1 gram $Ca(OH)_2$ in an Erlenmeyer flask using a magnetic mixer (24 hours) or a shaking table (4 hours), followed by settling (30 minutes). After settling, a filtration system was used consisting of a polyvinyl chloride (PVC) column [3/4 in (dia) x 12 in (l)] fitted with a PVC coupling containing a stainless steel screen (40 mesh), 15 grams of fine sand and several layers of filter paper. Samples were taken at different stages of the sorption process and analyzed.

Two-step batch demulsification sorption and filtration solution cleanup. A 100 milliter aqueous solution containing approximately 5000 mg/L of pesticide was mixed with 2 grams sorbent and 1 gram $Ca(OH)_2$ in an Erlenmeyer flask for 4 hours. The solution was then placed into a 100 ml bulb columns containing 2 grams of prewet sorbent. Samples were taken at various stages during the sorption process and analyzed.

Solid state fermentation. Solid state fermentation studies were conducted in twelve 38 L Rubbermaid waste containers fitted with locking lids. The organic matrix consisted of a sphagnum peat moss-cornmeal-crushed limestone mixture (67:22:11), dry wt basis. Bioreactors were inoculated with aged horse manure, agricultural soil and activated peat. The activated peat was collected from 5-year old field bioreactors previously used in pesticide degradation experiments. Addition of atrazine and carbofuran was accomplished using a 7.6 L insecticide pressure sprayer. Disappearance of pesticide residue and the appearance and disappearance of degradates from the organic matrix (hydroxyatrazine and carbofuran phenol) was monitored using high performance liquid chromatography.

Results and Discussion

Pesticide removal from aqueous solutions using one- and two step demulsification/sorption/filtration. A summary of results obtained from studies on pesticides formulated as emulsfiable concentrates, wettable powders or flowables treated by a one-step demulsification, sorption and filtration process is provided in Figure 1. Ten pesticide formulations contained in solutions approximating 5000 mg/L were mixed with one of three sorbent materials in the presence of a demulsification agent. The percentages of pesticide remaining in solution were calculated as changes from the initial

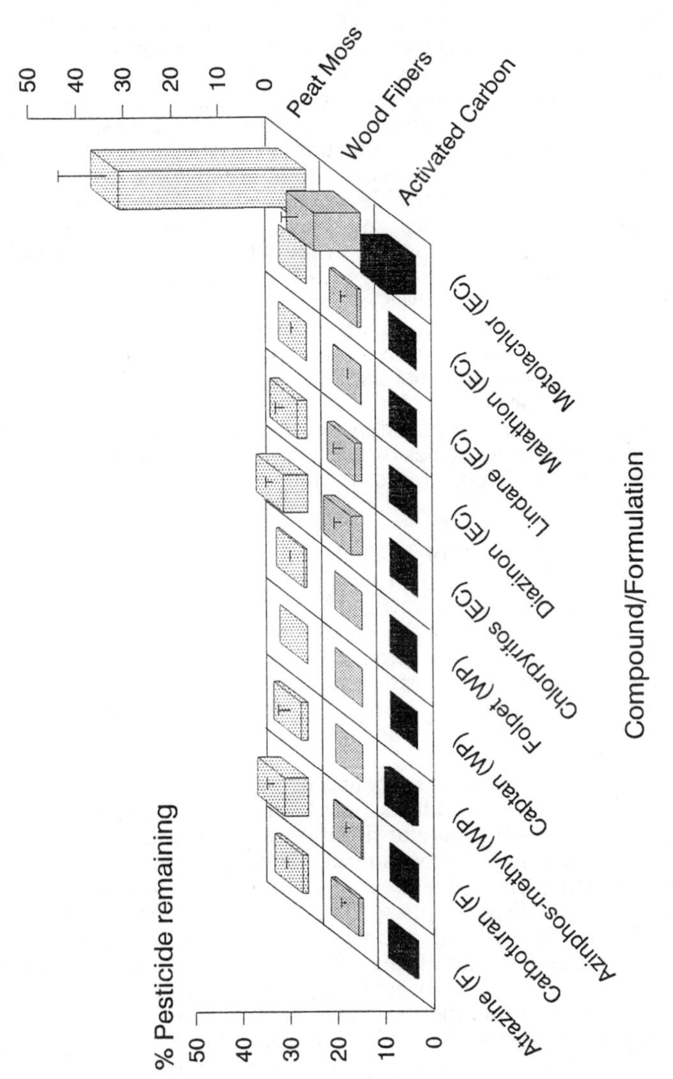

Figure 1. Efficiency of pesticide removal from aqueous suspension of various pesticide formulations using one-step demulsification, sorption and filtration.

concentration after the one-step treatment. The values provided are based on the mean of three replicate samples and quantitation limits were set at 1 mg/kg. Formulations were as follows: atrazine as AAtrex (50% ai), azinphos-methyl as Guthion (35% ai), captan (49% ai), carbofuran as Furadan (40.6%), chlorpyrifos as Dursban (44.4%), diazinon (48% ai), folpet (50% ai) lindane as Ortho (20% ai,) malathion as Dragon (50% ai) and metolachlor as Dual (86.4% ai). Emulsifiable concentrates (EC), flowables (F) and wettable powders (WP). Values expressed as the means ± standard error; error bars not shown when mean values or the errors were below 0 %. In most cases, large reductions in the pesticide concentration were achieved. It should be noted that sorption may not be the primary factor in removing the pesticide. Two other factors may contribute to the reduction in pesticide levels in these pesticide-laden solutions during this process. These are alkaline hydrolysis and filtration. Certain pesticides are unstable when they are exposed to alkaline conditions. The large reductions of captan, folpet and malathion concentrations were likely due to exposure to the alkaline conditions provided by using $Ca(OH)_2$ as a demulsification agent (4). We have not attempted to evaluate the extent of alkaline hydrolysis of these compounds which may occur during this process. However, extraction and analysis of sorbents used in these sorption studies revealed that in the case of these pesticides, little of the parent compound remained. Filtration of particulate suspensions provides for physical removal of pesticides from solutions containing wettable powders and flowable materials. Although the one-step treatment provides for reasonably good removal of most of the pesticides tested, improvements in the process are obvious. This is clearly demonstrated in the case of metolachlor and chlorpyrifos.

We envision that acceptable levels of pesticide removal provided by developing technology should be in the low mg/L (parts per million) range with a longer term goal of producing reductions in the low ng/L (parts per billion) range. As a result, we have initiated studies designed to examine the utility of including an additional step in the treatment procedure. A second step, employing column sorption and filtration has been added to the procedure (Figure 2). This process involves two phases: batch demulsification/sorption phase, and a disposal phase. The sorbent phase includes 1) batch demulsification and sorption where pesticide-laden waste solutions (or suspensions) are mixed with organic sorbents (lignocellulosic materials such as peat moss, processed wood products, etc.) and demulsification agents [$Ca(OH)_2$], and 2) column sorption and filtration where the solution is passed through a column containing lignocellulosic sorbent. During this phase, pesticides are removed from the aqueous solution by demulsification and sorption processes. The disposal phase begins after the separation of the sorbed pesticide from the treated aqueous solution by filtration (column sorption step). The aqueous solution may then be discarded as contaminant-free water and the pesticide-laden sorbent placed into bioreactors. Microbial populations are used to degrade the pesticide. Sorbents used in these studies have also been modified by amending them with vegetable oil. We have found pretreatment with vegetable oil provides better sorptivity of certain pesticides onto specific lignocellulosic matrices and is capable of enhancing the bioreactivity (rate of oxidative biological activity) of sorbents. Treatment of solutions containing either atrazine, chlorpyrifos or metolachlor using a two-step process with different sorbents improved pesticide removal (Table 1). In most cases, more than 99% of the original pesticide levels were removed. It can also be seen in this table that the addition of vegetable oil to steam-exploded wood fibers significantly improved the rate of metolachlor removal.

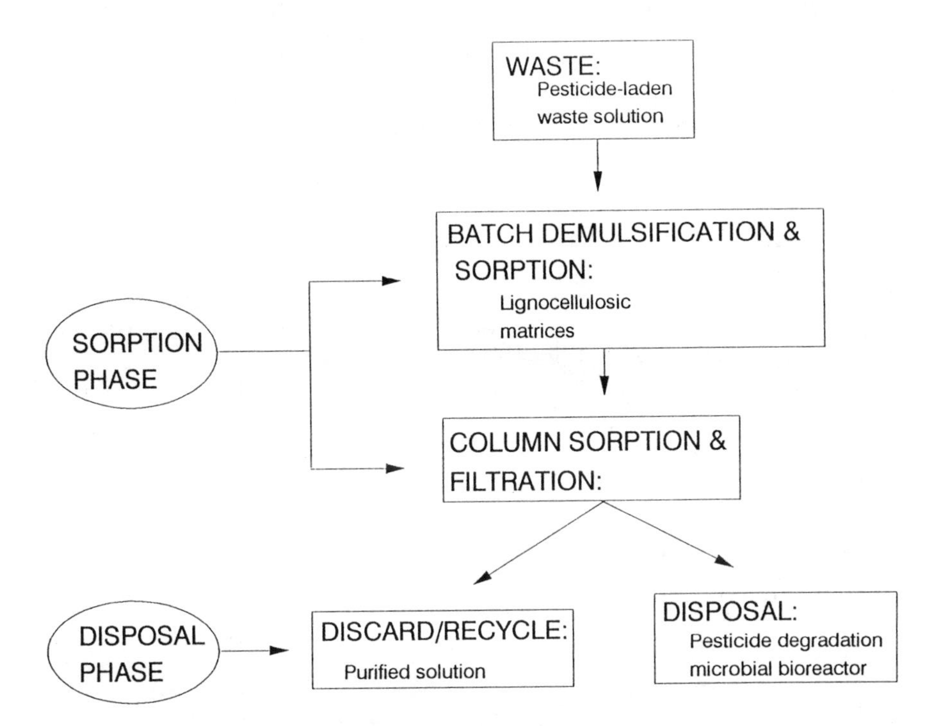

Figure 2. Model for pesticide wastewater disposal using organic sorption and microbial degradation.

Table 1. Removal of Several Pesticides from Aqueous Solutions Using Demulsification and Sorption as a Two-Step Process[a]

Pesticide[b]	Sorbent	Concentration mg/L[c]	% of Original concentration remaining
Atrazine	peat moss + 0% oil	14.7 ± 3.3 A a	0.3 ± 0.07
	peat moss + 10% oil	14.0 ± 12 a	0.3 ± 0.26
	wood fiber + 0% oil	52.3 ± 5.7 B a	1.14 ± 0.12
	wood fiber + 10% oil	48.7 ± 6.7 a	1.06 ± 0.15
Chlorpyrifos	peat moss + 0% oil	0.43 ± 0.09 A a	0.01 ± 0.0001
	peat moss + 10% oil	0.43 ± 0.23 a	0.01 ± 0.005
	wood fiber + 0% oil	21.0 ± 8.0 A a	0.46 ± 0.17
	wood fiber + 10% oil	27.3 ± 12.8 a	0.60 ± 0.28
Metolachlor	peat moss + 0% oil	3.2 ± 1.6 A a	0.07 ± 0.03
	peat moss + 10% oil	3.7 ± 0.9 a	0.08 ± 0.02
	wood fiber + 0% oil	120 ± 39 A b	2.6 ± 0.85
	wood fiber + 10% oil	2.3 ± 0.3 a	0.05 ± 0.01

[a] First step: demulsification and batch sorption; Second step: column sorption

[b] Pesticides: atrazine as **AAtrex** 4L, initial concentration = 4581 ± 875 mg/L; chlorpyrifos as **Dursban** 4E, initial concentration = 4584 ± 140 mg/L metolachlor as **Dual** 4E, initial concentration = 6096 ± 391 mg/L

[c] Statistical Comparisons: Capital letters compare sorbent types; sorbent type followed by the same capital letters are not significantly different ($P > 0.05$, coanalysis of variance). Lower case letters compare oil treatments of the sorbents; oil treatment (0/10% oil) oil treatments followed by the same lower case letter are not significantly different ($P > 0.05$, Studentized T - test). Values expressed as the means ± standard error of 3 replicate samples.

Solid State Fermentation

Once the solids have been removed from the aqueous phase by sorption and filtration (Figure 2), they will be placed in bioreactors to be degraded during the solid state fermentation process. We have conducted experiments on chemical and biological degradation of pesticides in bioreactors at several levels. We have used a benchtop composting system to examine the rate and metabolic fate of chlordane and diazinon using radiolabeled materials (**12**). Large bioreactors (6.7 cu ft) have been used to demonstrate efficacy of biodegradation of diazinon, chlorpyrifos, metolachlor, atrazine, carbofuran and chlordane (**5,7**). Some compounds are degraded quite rapidly in bioreactors containing peat moss enriched with corn meal. High levels of Diazinon AG500 (7.4 liters; at an estimated 66,000 mg/kg) were not detectable one year after the last addition of diazinon (**5**). Preliminary results from studies using 38 liter bioreactors containing peat moss amended with cornmeal to examine degradation of carbofuran and atrazine using enrichment microbial cultures are provided in Table 2. Carbofuran was degraded to undetectable levels. Although after 3 weeks of incubation the bioreactor medium contained 1.3% of the initial carbofuran as carbofuran phenol (a carbofuran hydrolysis product), it was undetectable after seven weeks. Degradation of atrazine sorbed onto nutrient-enriched peat moss resulted in significant concentration reduction (Table 2). After 26 weeks, only 14 % of the atrazine remained in the sorbent. After 7 weeks, almost 10 % of the initial concentration was present as hydroxyatrazine. After 15 weeks, hydroxyatrazine was no longer detectable. The disappearance of pesticide metabolites which appear during the fermentation process, supports the assumption that the parent molecule undergoes further degradation as the composting process progresses (**7**).

Summary

Information provided from our work supports the model which we have proposed for pesticide wastewater disposal using sorption and biodegradation. We have demonstrated that one-step demulsification, sorption and filtration provides effective removal of pesticides having low water solubilities. The pesticides tested represented several types of pesticides and formulations. Experiments with a two-step demulsification-sorption and sorption-filtration processes indicate that effective wastewater cleanup can be achieved in this manner.

Results which we have obtained compare favorably with reports of other work. For example, Dennis and Kolbyinsky (**2**) were able to remove 7 pesticides (100 mg/kg of each pesticide) contained in 400 gallons using 45 lbs of Calgon F-300 activated carbon in a Carbolator. After a 21 hour treatment, pesticide concentrations ranged from 0.5 to 5.6 mg/kg). Somich et al., (**14**). reported that after treatment of a mixture of 4 pesticides (ranging from 17 to 82 mg/kg) using ozone and biologically active soil columns, 1 to 20 % of the pesticides remained. Much higher initial concentrations (approximately 5000 mg/kg) have been used in our experiments, resulting in removal of significant quantities of the pesticides from the aqueous phase.

The advantages of this system include its relative simplicity, reliance on low cost materials, and safety. It is possible that the batch demulsification, sorption

Table 2. Degradation of Atrazine and Carbofuran in 38 Liter Bioreactors Containing Energy Enriched Peat Moss

Pesticide[a]/Degradation Metabolite	Concentration mg/kg				
	Initial	3 Weeks	7 Weeks	15 Weeks	26 Weeks
Atrazine	7231 ± 35	6901 ± 1074	6310 ± 951	5130 ± 962	980 ± 61
Hydroxyatrazine	0.0 ± 0.0	182 ± 33	525 ± 399	0.0 ± 0.0	
Carbofuran	749 ± 29	129 ± 38	0.0 ± 0.0		
Carbofuran phenol	0.0 ± 0.0	10 ± 15	0.0 ± 0.0		

[a] Pesticides: Atrazine as AAtrex 4L; Carbofuran as Furadan 4F; both applied as a liquid spray onto peat moss amended with in corn meal. Values expressed as the means ± standard error of 3 replicate samples.

and filtration step can be replaced by column demulsification, sorption and filtration. The use of columns could prove to be less cumbersome to users. Lignocellulosic sorbents proposed for use in the system show good promise because of their low cost and potential availability. Peat moss is readily available at low cost. Steam-exploded lignocellulosic materials represent a new sorbent resource that has shown good potential in our experiments. A variety of lignocellulosic materials (wood products, crop residues, recycled newsprint/paper) can be steam-exploded. Although these products are not yet commercially available, the estimated cost of bulk crude or unprocessed steam-exploded materials would be similar to that of peat moss.

One major advantage of the sorption disposal process which we propose is that once the rinsate has been treated, the sorbed pesticide no longer represents a major threat to the environment. If spilled, it can be collected and moved to a bioreactor quite easily. Should this system find broad application, and if the availability of pesticide-degrading enrichment cultures developed for a variety of pesticides is increased, it is likely that these consortia could be used to degrade pesticides in contaminated soil. Effective **in situ**, on site bioremediation may be enhanced by amending contaminated soils with lignocellulosic materials and pesticide-degrading microbes.

Several important questions regarding the rinsate solution cleanup process need to be answered. 1). There is a need for establishment of acceptable pesticide levels contained in treated rinsates destined for release in sewer systems, land application, etc. 2). If alkaline demulsification is used as a part of the process, it will be necessary to neutralize the treated rinsate. Our laboratory studies have indicated that small amounts of dilute solutions (0.2N) of hydrochloric acid can neutralize the treated rinsate. Hydrochloric acid (as muriatic acid) is relatively inexpensive and available at most hardware stores. 3). The final disposition of spent or reacted compost material when it is removed from bioreactors is another concern. Although the volume of lignocellulosic material will be reduced by as much as 20%, depending upon its nature during the composting process, most of it will remain. However, if pesticides are effectively degraded in the bioreactors, there will be little pesticide residue remaining in the spent matrix. We propose that if this is the case, this material could be land farmed, incinerated or possibly reused as a sorbent for additional waste water cleanup or **in situ** contaminated soil site remediation. We have several studies planned to address and evaluate these and other options which might be available for final disposal of these materials.

Based on these findings, we are continuing to develop this technology for field implementation and demonstration. Successful completion of this work should provide an acceptable, practical, effective, safe and inexpensive method for dilute and concentrated pesticide waste disposal.

Acknowledgements

This work was supported in part, by the Southern Region Impact Assessment Program, Virginia Center for Innovative Technology, USDA and the USEPA. Appreciation is extended to Ms. Andrea DeArment and Mr. Richard A. Tomkinson for the technical assistance they provided during major portions of our work.

References

1. Bridges, J. S; Dempsey, C. R. Eds. **Pesticide Waste Disposal Technology**; Noyes Data Corporation. Park Ridge, N J, 1988.
2. Dennis, W. H.; Kobylinski, E. A. *J. Environ. Sci. & Health* 1983, **18**, 317-331.
3. Hart, S. A. *In*: **Biological Processes, Innovative Hazardous Waste Treatment Series**; F. R. Sfrerra, Ed. Technomic Publishing Co. Lancaster, PA. 1991, 3, 7-17.
4. Hartley, D.; Kidd, H. Eds. **The Agrichemicals Handbook.** Royal Soc. Chem., The University of Nottingham, England. 1987, 2nd edition.
5. Hetzel, G. H.; Mullins, D. E.; Young, R. W.; Simonds, J. M. *In*: **Pesticides in Terrestrial and Aquatic Environments.** Proc. Nat. Res. Conf. Weighman, D. L. Ed.; Richmond, VA, 1989, pp 239-248.
6. Judge, D. N.; Mullins, D. E.; Hetzel, G. H.; Young, R. W. *In:* **Pesticides in the Next Decade: The Challenges Ahead.** Proc. 3rd Nat. Res. Conf. on Pesticides, Weighman, D. L. Ed.; Richmond, VA, 1990, p 145-158.
7. Mullins, D. E.; Young, R. W.; Palmer, C. P.; Hamilton, R. L.; Sherertz, P. C. *Pestic. Sci.* 1989, **25**: 241-254.
8. Mullins, D. E.; Young, R. W.; Hetzel G. H.; Berry, D. F. *In* **Proceedings of the International Workshop on Research in Pesticide Treatment/ Disposal/ Waste Minimization**, Ferguson, T. D. Ed.; USEPA, Cincinnati, OH, **EPA/60019-91047. 1992, pp.32-45.**
9. Norwood, V. M. **A Literature Review of Waste Treatment Technologies Which May be Applicable to Wastes Generated at Fertilizer-Agrichemical Dealer Sites.** Tennessee Valley Authority Bulletin Y-214 Muscle Shoals, AL, 1990, 32pp.
10. Overend, R. P.; Chornet, E. *Phil. Trans. R. Soc. Lond. A*. 1987,**321**, 523-536.
11. Parr, J. F.; Epstein, E.; Willson, G. B. *Agri. Environm.* 1978, **4**, 123- 137.
12. Petruska, J. A.; Mullins, D. E.; Young, R. W.; Collins, E. R. *Nuclear and Chem. Waste Manag.* 1985, **5**, 177-182.
13. Seaman, D. *Pestic. Sci.* 1990, **29**, 437-449.
14. Somich, C. J.; Muldoon, M. T.; Kearney, P. C. *Environ. Sci Technol.* 1990, **24**, 745-749.
15. Thomas, B. *Pestic. Sci.* 1990, **29**, 475-479.
16. Ware, G. **The Pesticide Book.** Thomson Publications, Fresno, CA, 1989, 3rd ed. 340pp.
17. Watts, R. R. **Analytical Reference Standards and Supplemental Data for Pesticides and Other Organic Compounds.** 1981, USEPA-600-12-81-001.
18. Williams, R. T.; Myler, C. A. BioCycle. 1990, **31**, 78-82.

RECEIVED June 1, 1992

Chapter 15

Evaluation of Organophosphorus Insecticide Hydrolysis by Conventional Means and Reactive Ion Exchange

Kathryn C. Dowling and Ann T. Lemley[1]

Graduate Field of Environmental Toxicology, Cornell University, Ithaca, NY 14853

Different methods of hydrolyzing the organophosphorus insecticide methyl parathion were compared for effectiveness. The aqueous base hydrolysis rate is second order in the insecticide and sodium hydroxide. In a system incorporating sodium perborate, hydrolysis rates are accelerated by two orders of magnitude in comparison with simple aqueous base hydrolysis. Pseudo-first order rates are linear with sodium perborate concentrations for a given sodium hydroxide concentration. A macroporous hydroxide-presenting resin employed in methyl parathion batch studies catalyzes hydrolysis at somewhat less than the rate of simple base hydrolysis. Reactive ion exchange with this resin in a dynamic flow-through column system degrades four organophosphates: methyl parathion, malathion, chlorpyrifos, and methamidophos. Experimental solutions prepared in tap water instead of distilled water are significantly less degraded (intact insecticide appeared in the column effluent). Ions present in tap water may interfere with resin/insecticide interactions, decreasing degradative capacity.

The need for development of pesticide treatment systems easily accessible for field use stems from the large amount of insecticide application equipment, including mixing apparati, sprayers, and fumigation airplanes, in use today. Pesticide-contaminated rinsates generated from emptying and cleaning such equipment can contain pesticide concentrations in the range of 100 to 1000 mg L^{-1} (*1*). These rinsates, often quite toxic, are not always collected for further use or treatment. Since discarded rinsate can contaminate surface waters and groundwater, field-accessible treatment systems are desirable. A number of hydrolysis treatment schemes that ultimately could be engineered for field use were evaluated for their ability to degrade organophosphorus insecticides. Four organophosphates widely used in

[1]Corresponding author. Current address: College of Human Ecology, 202 MVR Hall, Cornell University, Ithaca, NY 14853–4401

0097–6156/92/0510–0177$06.00/0

agriculture were chosen for study: methyl parathion, malathion, methamidophos, and chlorpyrifos. In an attempt to represent actual field conditions, experiments were carried out in aqueous solution at concentrations reflecting insecticide water solubilities and common rinsate levels.

The first method investigated, aqueous hydrolysis by hydroxide, is often used to degrade organophosphates. Hydroxide nucleophilic attack on the central organophosphate phosphoester leads to insecticide cleavage. Hydrolysis rate constants can be calculated easily for a range of pH values. Most previous work on organophosphate base hydrolysis has been done in organic solvents, but one aqueous study can be compared to the present work. Ketelaar (*2*) found methyl parathion's second order rate constant (15°C) to be 9.2×10^{-2} M^{-1} min^{-1} and its temperature coefficient (for each 10°C increment) to be 2.57. Thus the expected hydrolysis rate constant at 25°C is 2.36×10^{-1} M^{-1} min^{-1}. Organophosphates are also susceptible to hydrolysis by the perhydroxyl anion, a fifty-fold stronger nucleophile than the hydroxyl anion (*3*,*4*). Sodium perborate forms the perhydroxyl anion and boric acid in water under alkaline conditions. Higher pH values favor perhydroxyl anion formation, which in turn heightens hydrolysis rates. Boric acid is slightly toxic to mammals (with LD_{50} values of 1000-6000 mg kg^{-1} in various species); boron, although commonly present in water and soil and essential to plant growth, is of ecological concern and has a drinking water concentration limit of 1 mg L^{-1} (*5*).

Janauer *et al.* (*6*) and Lemley *et al.* (*7*) modified various ion exchange resins and gels by loading with hydroxyl or other ions; degradation was effected by passing aqueous solutions of organophosphate or carbamate compounds through a resin-packed column. The hydroxide-modified resin was found to be particularly effective at degrading organophosphates. The insecticide hydrolysis product was found to be retained in its anionic form by the resin; this was a welcome concentration effect for degradation products. Products could be discarded through resin elution/regeneration or disposed of while still bound to the resin. Hydroxyl ion, perhydroxyl ion, and reactive ion exchange (RIEX) hydrolysis of organophosphates will be described and evaluated.

Materials and Methods

Chemicals. Methyl parathion, *O*,*O*-dimethyl *O*-4-(nitrophenyl) phosphorothioate (98.0% labeled/98.0% confirmed purity), was purchased from Chem Service, Inc., West Chester, PA. Chlorpyrifos, *O*,*O*-diethyl *O*-(3,5,6-trichloro-2-pyridyl) phosphorothioate (99.9% labeled/100.0% confirmed purity), along with its hydrolysis product 3,5,6-trichloro-2-pyridinol (99% labeled/100% tested purity), was donated by Dow Chemical Company, Midland, MI. Malathion, *O*,*O*-dimethyl *S*-1,2-bis(ethoxycarbonyl) ethyl phosphorodithioate (98.4% labeled/90.6% actual purity), was given by American Cyanimid Company, Princeton, NJ. The methyl parathion and malathion hydrolysis products *p*-nitrophenol (>99% labeled/98.8% confirmed purity) and diethyl fumarate (98% labeled/100% actual purity) were purchased from Aldrich Chemical Company, Milwaukee, WI. Methamidophos, *O*,*S*-dimethyl phosphoramidothioate (95.8% labeled/91.1% confirmed purity), was contributed by Mobay Corporation, Kansas City, MO. The external standard, ethoprop, *O*-ethyl

S,S– dipropyl phosphorodithioate (99.7% labeled/100% tested purity), donated by Rhone-Poulenc Ag Company, Research Triangle Park, NC, was used for all GC/ MSD analyses.

HPLC grade methanol, reagent grade nitric acid, and certified 85% *o*-phosphoric acid were purchased from Fisher Scientific, Pittsburgh, PA. Sodium hydroxide and sodium perborate tetrahydrate were Mallinckrodt products (Paris, KY), and Aldrich Chemical Company, respectively. The analytical grade macroporous anion exchange resin (100-200 mesh) used, AG MP-1 in the chloride form, came from Bio-Rad Laboratories, Richmond, CA. Unless otherwise noted, aqueous solutions were prepared with distilled water (reverse osmosis water passed through a Barnstead NANOpure II 4-Module System and distilled in a Corning MEGA-PURE Model MP-1 Still). Some experiments utilized Cornell University tap water supplied by a nearby stream, fully filtered and chemically treated. When tested by silver titration on a Buchler-Cotlove chloridometer, the water was found to contain 25 mg L^{-1} Cl^{-}.

Analytical Methods. Methyl parathion, chlorpyrifos, malathion, and their decay products were analyzed by GC/MSD using a Hewlett Packard 5890 Series II Gas Chromatograph/Hewlett Packard 5971A Mass Selective Detector equipped with a Supelco SPB-608 Fused Silica Capillary Column (0.25 µm film, 0.25 mm ID x 30 m). GC conditions were as follows: splitless/split injector at 280°C for a 1µL injection in methanol; oven temperature programmed from 150°C to 250°C at 10°C/min and held for 2 min at 250°C; MS detector set at 280°C; and ultra-high purity helium carrier gas fixed at 1.2 mL min^{-1} flow rate. Detection by single-ion monitoring of each compound's most abundant ion maximized quantitative ability. Each compound was analyzed by GC/MSD in the scan mode to confirm purities. In some cases purities differed from labeled purities; impurities were identified, quantified and used to establish actual purities (see Chemicals above for confirmed test purities). Experimental insecticide decay was quantified with parent insecticide disappearance data. Methyl parathion and chlorpyrifos break-down products' identities and presence were confirmed for each experiment to show that hydrolysis had occurred. The malathion hydrolysis product under aqueous alkaline conditions, diethyl fumarate (*8,9*), was identified, but malathion monoacid was not detected.

Solid phase extraction was used to transfer insecticides from aqueous solution to methanol. Bakerbond cyclohexyl SPE columns were conditioned with 3 mL methanol then 5 mL distilled water. Ten milliliters of insecticide solution was acidified with 1 *M* HNO_3 to pH values appropriate for converting each break-down product (as a weak acid) to its unionized form. These solutions were passed through the cartridges which were dried and eluted with 4 mL of methanol. One mL external standard solution (82 or 83 mg L^{-1} ethoprop in methanol) was added and the solution was brought up to 5 mL volume and analyzed by GC/MSD. Quantitation against the ethoprop standard normalized instrumental response.

The high water solubility of methamidophos made ineffective its solid phase extraction for GC/MSD; thus it was analyzed using a Hewlett-Packard 1090A HPLC equipped with a Supelco LC-8-DB column (3µm particle size 4.6 mm I.D. x 15 cm) and a Supelguard LC-8-DB guard column. Injections of 200 µL sample in water were made on a 200 μL sample loop; oven temperature was 50°C; the mobile phase was 10%/90% methanol/water at 1 mL min^{-1}; and detection by diode-array detector was at 212 nm. Quistad *et al.* (*10*) identified the methyl thiolate ion as the species liberated during methamidophos hydrolysis. No methamidophos break-down products were detected by HPLC, but the pungent mercapturic odor released during methamidophos experiments pointed to methyl mercaptan evolution.

To obtain rapid information on methyl parathion kinetic behavior, a spectrophotometric method was developed. Formation of the methyl parathion alkaline hydrolysis product, the bright yellow *p*-nitrophenolate ion, was followed at 400 nm on a Perkin-Elmer Lambda 2 Ultraviolet/Visible Spectrometer. The molar absorptivity of *p*-nitrophenol was measured as 18050 M^{-1} cm^{-1}.

Experimental Methods

Hydroxyl Ion Hydrolysis. Aqueous mixtures of 1.96 x 10^{-4} *M* methyl parathion plus separate hydroxide solutions of 0.0125, 0.025, 0.0375, 0.05, 0.075, and 0.100 *M* were prepared in quartz cells. The formation of the *p*-nitrophenolate ion was monitored spectroscopically; absorbance versus time curves for each sodium hydroxide concentration were generated. All absorbance values were converted to *p*-nitrophenol concentrations, [*p*NP], using *p*-nitrophenol's molar absorptivity. Methyl parathion concentrations, [MP], were calculated at a given time t, assuming *p*-nitrophenol production in a one-to-one molar ratio with methyl parathion destruction:

$$[MP]_t = [MP]_{t=o} - [pNP]_t \quad (1).$$

With sodium hydroxide present in excess, the expected pseudo-first order reaction rate constant, k_{obs}, is represented by the slope of the plot of the natural log of the fraction of methyl parathion remaining at time t versus t:

$$k_{obs} = \ln ([MP]_t / [MP]_{t=0}) / t \quad (2).$$

In turn, a linear regression of k_{obs} values versus hydroxide concentrations yields a slope equivalent to the second order rate constant, k_r, as shown. Pseudo-first order reaction rate constants were calculated and used to determine the second order rate constant.

$$k_r = k_{obs} / [OH^-] \quad (3).$$

Perhydroxyl Ion Hydrolysis. Perhydroxyl ion hydrolysis experiments were conducted at several concentrations of both sodium perborate, sodium hydroxide, and methyl parathion. These experimental conditions are given in Table I. Spectrophotometric analysis of the *p*-nitrophenolate product and calculation of first order

Table I. Observed Hydrolysis Rates of Methyl Parathion in the Presence of Perhydroxyl Ion

[OH]*	[PB]*	[MP]*	k_{obs} (min^{-1})	R^2	$t_{1/2}$ (min)
9.32×10^{-6}	1.30×10^{-3}	1.31×10^{-4}	5.1×10^{-3}	0.999	136
	2.61×10^{-3}		1.2×10^{-2}		58
	6.52×10^{-3}		3.9×10^{-2}		18
	1.30×10^{-2}		8.5×10^{-2}		8
1.25×10^{-5}	1.95×10^{-3}	1.95×10^{-4}	7.6×10^{-3}	0.989	91
	3.90×10^{-3}		1.7×10^{-2}		41
	9.75×10^{-3}		6.0×10^{-2}		12
	1.95×10^{-2}		1.0×10^{-1}		7
9.32×10^{-5}	1.30×10^{-3}	1.31×10^{-4}	6.5×10^{-3}	0.999	104
	2.61×10^{-3}		1.5×10^{-2}		46
	6.52×10^{-3}		4.0×10^{-2}		17
	1.30×10^{-2}		8.6×10^{-2}		8
1.25×10^{-4}	1.95×10^{-3}	1.95×10^{-4}	9.8×10^{-3}	1.000	71
	3.90×10^{-3}		2.1×10^{-2}		33
	9.75×10^{-3}		6.0×10^{-2}		12
	1.95×10^{-2}		1.2×10^{-1}		6
8.47×10^{-4}	1.27×10^{-3}	1.31×10^{-4}	1.4×10^{-2}	0.995	50
	2.61×10^{-3}		1.9×10^{-2}		36
	6.52×10^{-3}		3.9×10^{-2}		18
	1.31×10^{-2}		6.3×10^{-2}		11
1.25×10^{-3}	1.95×10^{-3}	1.95×10^{-4}	3.9×10^{-2}	0.95	18
	3.90×10^{-3}		5.9×10^{-2}		12
	9.75×10^{-3}		1.2×10^{-1}		6
	1.95×10^{-2}		1.6×10^{-1}		4
6.68×10^{-3}	1.27×10^{-3}	1.31×10^{-4}	7.8×10^{-2}	0.999	9
	2.61×10^{-3}		1.1×10^{-1}		6
	6.52×10^{-3}		1.8×10^{-1}		4
	1.31×10^{-2}		3.0×10^{-1}		2
1.25×10^{-2}	1.95×10^{-3}	1.95×10^{-4}	1.7×10^{-1}	0.972	4
	3.90×10^{-3}		2.7×10^{-1}		3
	9.75×10^{-3}		4.7×10^{-1}		2

*[OH], [PB], and [MP] represent the molar concentrations of sodium hydroxide, sodium perborate and methyl parathion, respectively.

observed reaction rate constants were accomplished as described in the section above.

RIEX Batch Experiments. Janauer *et al.* (*6*) approached methyl parathion batch studies by holding the resin amount constant at 0.3 g, applying known amounts of methyl parathion aqueous solution, and varying batch incubation times from 5 to 60 minutes. In the current study, all batch studies were conducted for a constant period of time, 25 hours, at ambient temperature (23-27°C) and with varying amounts of resin. In all, four sets of batch studies were run: two pairs of distilled water and two pairs of laboratory tap water for two different solution volumes.

Each study set consisted of five different resin quantities, 0.2, 0.4, 0.6, 0.8, and 1.0 g. The AG MP-1 resin was activated to the hydroxide form in a glass column (10 mm I.D. x 10 cm long) supported by a fritted glass disc. Resin was rinsed into columns with a generous amount of distilled water and twenty resin volumes of 1 *M* NaOH solution were passed over it. The resin was then rinsed with twenty volumes of distilled water (at least 8 mL). A qualitative silver nitrate test verified the absence of chloride ion (no white precipitate observed) and sufficiently neutral pH (no yellow precipitate present). Each resin sample was transferred into an Erlenmeyer flask. In the first pair of batch experiments, resin samples were added to 250-mL Erlenmeyer flasks with 50 mL of methyl parathion solution of 35.6 mg L^{-1} in distilled water or 50.1 mg L^{-1} in laboratory tap water. In the second pair, resins were added to 500-mL Erlenmeyer flasks with 200 mL of methyl parathion solution of 54.4 mg L^{-1} in distilled water or 39.4 mg L^{-1} in tap water.

All batch samples were agitated on an Eberbach shaker at low setting for 25 hours, after which they were filtered with Whatman No. 1 qualitative filter paper and the filtrates analyzed by GC/MSD. The filtered resin was eluted by passing over it 10 mL of a 1:1 mixture of H_3PO_4:distilled water. This solution was then diluted ten-fold and analyzed by GC/MSD to confirm the presence of *p*-nitrophenol.

RIEX Column Experiments. A dynamic flow system allowed examination of the effect of varying flow rate as well as solution concentrations, application volumes, resin amounts, and distilled versus tap water. AG MP-1 resin was prepared as described in the section above. At the start of an experiment, the water level was lowered to the top layer of resin so the only excess water remaining was that saturating the resin. A Sage Instruments dual syringe pump supplied insecticide treatment solution at a constant flow rate to the resin. A (~3 mL) reservoir was established and maintained on top of the resin. A hose clamp assembly at the bottom of the column allowed simple gravity flow at a rate equal to the influx. An Isco Retriever II fraction collector periodically gathered samples for analysis over the course of the experiment. At the end of an experiment, solution was drained from the resin, and the resin was eluted by passing over it 5 mL of H_3PO_4 and 5 mL distilled water. This solution was diluted 25-fold and analyzed by GC/MSD to confirm the presence of insecticide break-down product.

Results and Discussion

Hydroxyl Ion Hydrolysis. Plots of ln [MP] versus time are linear (all R^2 values were ≥ 0.996). Values of k_{obs} range from 3.88×10^{-3} to 2.37×10^{-2} for the sodium hydroxide concentrations used (0.0125 to 0.10 M); corresponding half-life values ($t_{1/2} = 0.693 / k_{obs}$) are 3.0 to 0.5 hours. The plot of observed reaction rate constants versus $[OH^-]$ is shown in Figure 1. The k_r value for methyl parathion in the presence of excess hydroxide at 25°C, found to be 0.246 M^{-1} min^{-1} ($R^2 = 0.989$), is close to the value of 0.236 M^{-1} min^{-1} predicted by Ketelaar (*2*). This rate constant is nearly 400-fold smaller than that (also at 25°C) of 91.5 M^{-1} min^{-1} (*11*) observed for aldicarb sulfone base hydrolysis (solutions in the part per million range). Aqueous base hydrolysis data provide a basis for comparison with other hydrolysis treatment methods. Methyl parathion is easily and rapidly hydrolyzed in solutions of 50- to 500-fold excess of hydroxide. The reaction is second order and dependent on both insecticide and base. The major disadvantage of the system is the high pH values needed to attain high hydrolysis rates.

Hydroxyl/Perhydroxyl Ion Hydrolysis. Observed reaction rate constants (k_{obs}) in the presence of perhydroxyl ion are given in Table I. These constants can be compared with those for base hydrolysis at the sodium hydroxide concentration of 0.0125 *N*. This concentration was the lowest examined for hydroxyl but the highest for perhydroxyl ion hydrolysis. The half-life value in the former case is 3.0 hours; in the latter case values range from 4 to 1.5 minutes, depending on the sodium perborate concentration. Thus, reaction rates are accelerated 45, 60, or 120-fold simply by adding 10-,20-, and 50-fold excess amounts of sodium perborate over insecticide. It is important to note that the addition of a ten- or twenty-fold excess of perborate ion causes significant acceleration of methyl parathion hydrolysis at pH values above eleven. Conversely, perborate ion significantly lowers the amount of sodium hydroxide required to produce a given k_{obs}. The k_{obs} with a 50-fold perborate excess at pH 9 (see the third line of Table I) is higher than the k_{obs} with no perborate at pH 13 (2.37×10^{-2} min^{-1}).

Batch RIEX Studies. Batch results are shown in Table II. Of the twenty samples incubated in the batch system, 15 showed 100% methyl parathion degradation over the course of the study. These include 1) ten experiments using the five resin amounts diluted with distilled water to the two solution volumes (50 or 200 mL); 2) one experiment for the 1.0 g resin amount diluted to 200 mL with tap water, and 3) four experiments for 0.4, 0.6, 0.8, and 1.0 g resin diluted to 50 mL with tap water. In all cases, the decay product, *p*-nitrophenol, is found to be present only on the resin; no methyl parathion is detected in solution.

Since the remaining five samples, all diluted with tap water (see Table II for details), demonstrate incomplete decay, pseudo-first order rate constants were calculated (Equation 2) using length of experiment (25 hr) and percent methyl parathion remaining in solution. "Equivalent" hydroxide concentrations were calculated with the manufacturer-stipulated exchange capacity of 4.2 meq g^{-1} dry AG MP-1 resin. Assuming that all active sites were converted to the hydroxide form

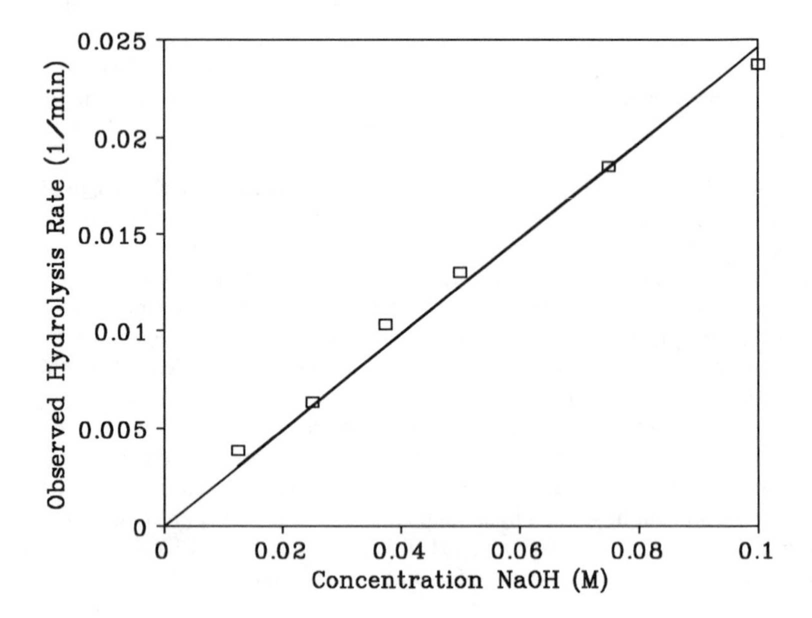

Figure 1. Effect of Hydroxide Ion Concentration on Methyl Parathion Disappearance

Table II. RIEX Batch Studies: Experimental Conditions and Results

Amount Resin (g)	Solution* Volume (mL)	[MP]	Equivalent [OH]	Percent Recovered	k_{obs} (min^{-1})	k_r ($M^{-1}min^{-1}$)
0.2	50	1.9×10^{-4}	1.7×10^{-2}	3.6	2.2×10^{-3}	0.13
0.2	200	1.5×10^{-4}	4.2×10^{-3}	74	2.0×10^{-4}	0.049
0.4	200	1.5×10^{-4}	8.4×10^{-3}	34	7.3×10^{-4}	0.087
0.6	200	1.5×10^{-4}	1.3×10^{-2}	14	1.3×10^{-3}	0.10
0.8	200	1.5×10^{-4}	1.7×10^{-2}	3.0	2.3×10^{-3}	0.14

*All samples listed were prepared with tap water.

allowed a calculation of the molar amount of hydroxide on the resin. This was divided by the solution volume to obtain an "equivalent" hydroxide concentration used to calculate a second order reaction rate constant, k_r (Equation 3).

Second order rate constants in the batch system for resin-mediated hydrolysis of methyl parathion in tap water solution range from 0.05 to 0.14 M^{-1} min^{-1} (Table II). Lower rate constants correspond to low resin:solution ratios and are likely due to the fact that hydroxide ions bound to the resin are not evenly distributed throughout the entire solution. The values are somewhat lower than that of 0.25 M^{-1} min^{-1} calculated for aqueous alkaline hydrolysis. The two rate constants represent very different situations: a batch system where hydroxyl ions are bound to a resin agitated in solution and a solution where hydroxyl ions are evenly distributed with respect to methyl parathion molecules. In summary, a reactive ion exchange resin exhibits a hydrolysis reaction rate nearly as high as does simple aqueous hydrolysis while maintaining lower solution pH values and binding undesirable products to the resin.

Column RIEX Studies. Column RIEX results are presented in Figures 2-6 as insecticide break-through curves. Relative effluent insecticide concentration (compared with the initial insecticide concentration applied to the column) is plotted against the relative amount of insecticide applied over the course of the study. In this way, both axes are normalized to values between zero and one, facilitating comparison of experiments with different conditions. Table III lists these conditions for each insecticide by letters that correspond to those on the figures.

Table III also includes the overall percent of insecticide degraded by the resin as well as the corresponding pseudo-first order observed hydrolysis rate constant, k_{obs}. This constant was calculated for each experiment in Table III based on the residence time of an individual insecticide molecule on the RIEX column. Manufacturer specifications for the AG MP-1 resin include a density, ρ, of 0.70 g cm^{-3} and a saturation void volume of 33% of the resin volume. Thus the resin amount (A) used was converted to residence time (t_R, min):

$$t_R = (A/\rho)\ (0.33/F) \quad (4),$$

where F is the measured experimental flow rate (mL min^{-1}). The residence time was used to calculate the first order reaction rate constant as in Equation 2 above. The rate constants obtained in this way are higher, by one to nearly four orders of magnitude, than those obtained for the batch system (Table II). The efficiency of the dynamic flow system in bringing together insecticide and hydroxyl ions is illustrated by these high rate constants.

Methyl Parathion. To examine the effect of varying experimental conditions, a number of methyl parathion break-through curves were generated by applying sufficient insecticide to exhaust the resin. Figure 2 demonstrates the effect of varying the flow rate across the resin and shows that higher flow rates result in less overall degradation (Table III; experiments A, B, and C). Although pseudo-first order hydrolysis rate constants (as calculated) should be proportional to flow rates,

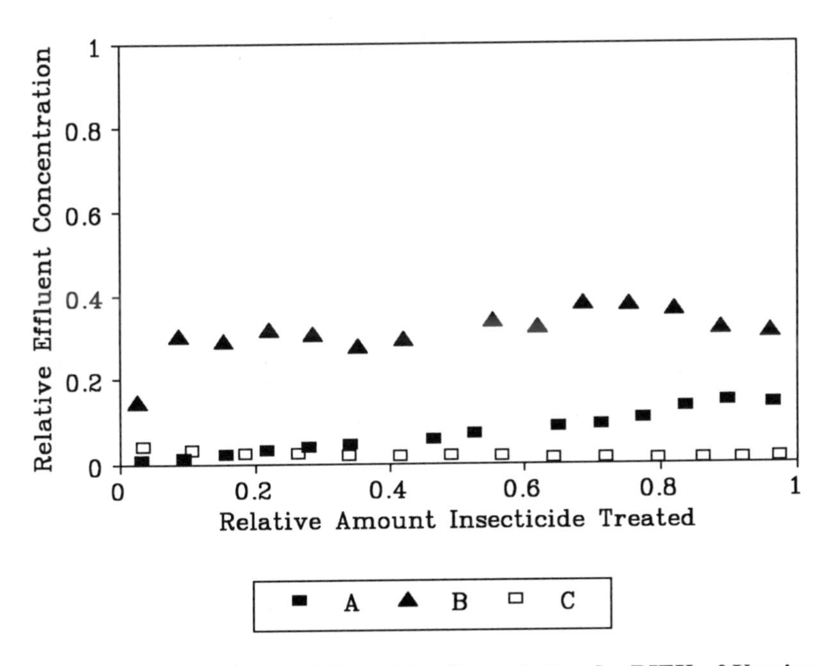

Figure 2. Effect on Methyl Parathion Degradation by RIEX of Varying Application Flow Rates

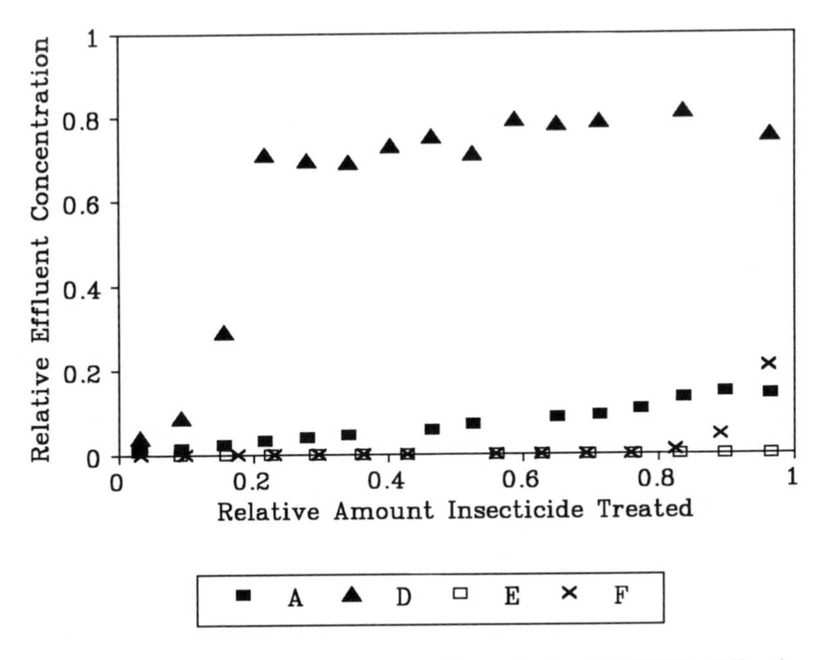

Figure 3. Effect on Methyl Parathion Degradation by RIEX of Varying Resin Amounts for Solutions of Distilled or Tap Water

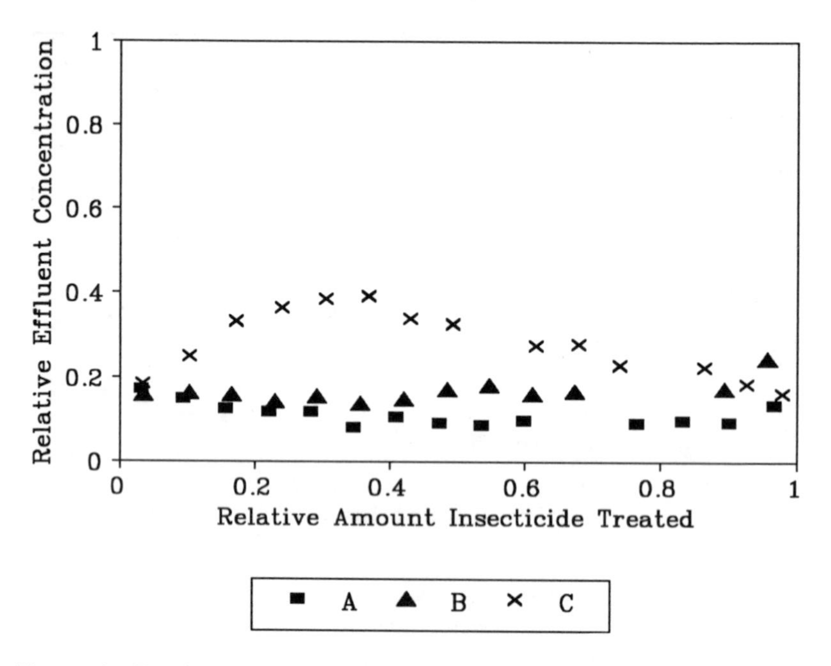

Figure 4. Degradation of Chlorpyrifos by RIEX Under Varying Experimental Conditions

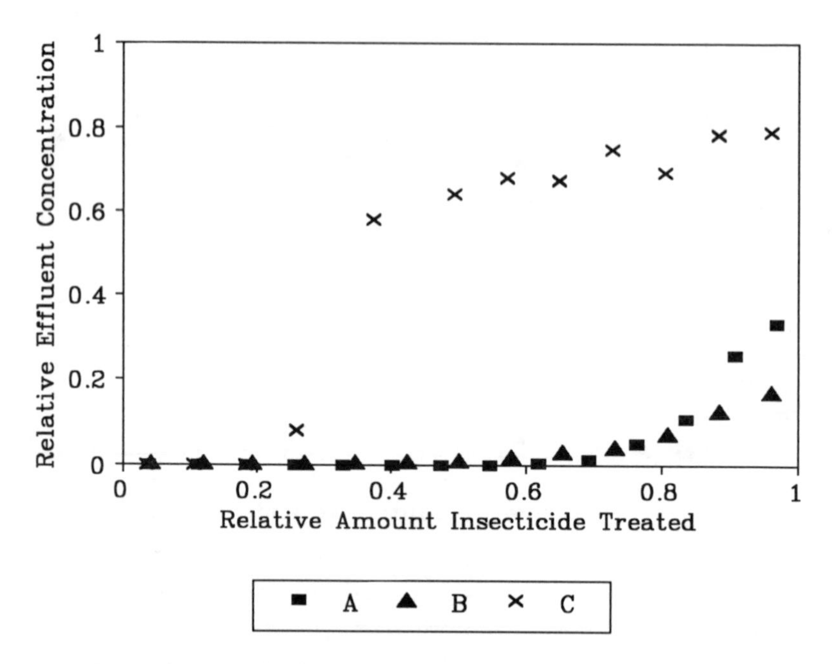

Figure 5. Degradation of Malathion by RIEX Under Varying Experimental Conditions

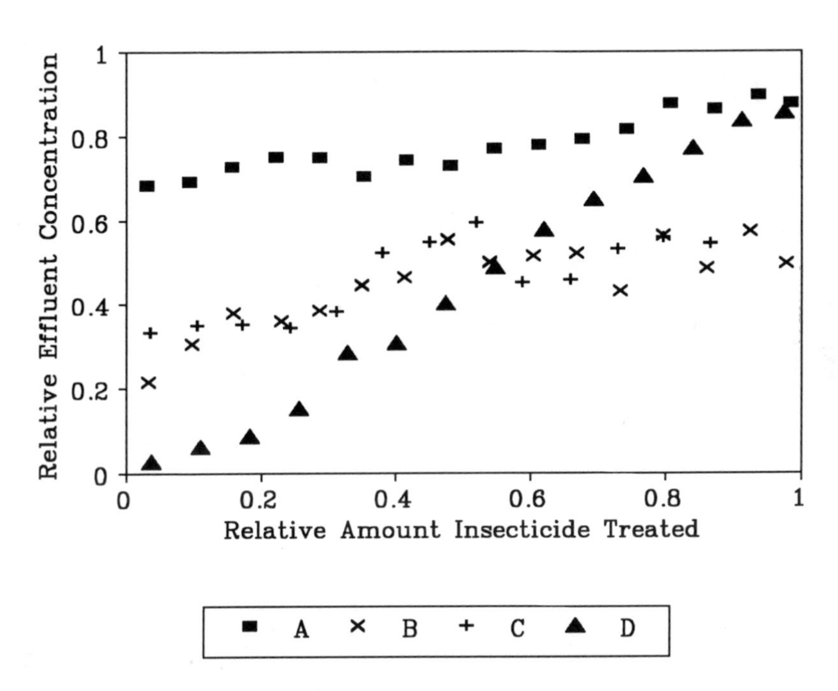

Figure 6. Degradation of Methamidophos by RIEX Under Varying Experimental Conditions

Table III. RIEX Column Studies: Experimental Conditions and Results

Exp.	Concentration Applied (M)	Volume Applied (mL)	Flow Rate (mL min^{-1})	Amount Resin (g)	Type of Water	Percent Recovered	k_{obs} (min^{-1})
Methyl Parathion							
A	1.5×10^{-4}	190	0.9	0.2	Dist.	7.1	26
B	1.8×10^{-4}	190	2.8	0.2	Dist.	31.4	35
C	1.6×10^{-4}	200	0.3	0.2	Dist.	2.0	11
D	1.7×10^{-4}	190	0.9	0.2	Tap	63.9	4.0
E	1.9×10^{-4}	190	0.9	1.0	Dist.	<0.1	15
F	1.7×10^{-4}	190	0.9	1.0	Tap	1.9	7.7
G	3.6×10^{-4}	190	0.9	1.0	Dist.	<0.1	*
H	8.2×10^{-4}	190	0.9	1.0	Dist.	<0.1	*
Chlorpyrifos							
A	1.4×10^{-4}	240	0.9	0.2	Dist.	9.1	22
B	1.6×10^{-4}	240	2.6	0.2	Dist.	16.7	49
C	1.5×10^{-4}	240	0.9	0.2	Tap	21.0	15
Malathion							
A	3.4×10^{-4}	150	0.9	0.2	Dist.	5.2	28
B	4.5×10^{-4}	148	2.8	0.2	Dist.	3.8	99
C	3.8×10^{-4}	141	0.9	0.2	Tap	54.1	5.9
Methamidophos							
A	2.8×10^{-3}	195	2.9	0.2	Dist.	67.9	12
B	2.9×10^{-3}	45	1.0	0.2	Dist.	45.0	8.1
C	6.9×10^{-4}	160	0.9	0.2	Dist.	46.1	7.6
D	3.3×10^{-3}	90	0.3	0.2	Dist.	43.4	2.4

*Rate constants could not be calculated since no untreated insecticide was recovered in the effluent.

they increase proportionately less than do flow rates, indicating that the flow (in the range studied) significantly affects methyl parathion degradation.

Figure 3 compares experiments using distilled versus tap water and small (0.2 g) versus large (1.0 g) amounts of resin. Tap water use decreases the amount of methyl parathion degraded; this effect is observed for both resin amounts studied, but more dramatically for the lower resin amount. For 0.2 g resin (A versus D), the observed rate constant is 6.5 times lower and for 1.0 g (E versus F), it is two times lower. The same effect is observed in the batch system and may be explained by the interference of ions present in tap water with ion exchange in the resin/insecticide system. Tap water anions, such as chloride, probably compete with hydroxide for resin active sites whereas cations compete with the insecticide for interaction with the bound anions.

Two additional methyl parathion experiments (G and H) using distilled water, flow rates of 0.9 mL min^{-1} and 1.0 g resin were performed with methyl parathion solutions of 3.65×10^{-4} and 8.24×10^{-4} *M* (1.9 and 4.3 times higher than methyl parathion's water solubility). No methyl parathion is found either in the column effluent or on the resin, demonstrating that the RIEX system is capable of degrading all insecticide applied given appropriate flow rates and resin amounts. Elution of resin by passing the $H_2PO_4^-$ anion across it yields the hydrolysis product *p*-nitrophenol, as in experiments A through F. Exhaustion of the resin is evidenced by undegraded methyl parathion break-through and appearance in the effluent (experiments A through F).

Chlorpyrifos. The chlorpyrifos hydrolysis product, 3,5,6-trichloro-2-pyridinol, like *p*-nitrophenol, is recovered from the resin and not found in solution; it is bound by the resin. Figure 4 shows that chlorpyrifos break-through curves have a somewhat different pattern, with less S-shaped character, than those of methyl parathion. When compared to each other, curves A and B differ less than do their equivalents (A and B: Figure 2) for methyl parathion. Thus, an equivalent change in flow rate has less of an effect on chlorpyrifos degradation than on methyl parathion degradation.

The tap water curve (C) initially shows increasing chlorpyrifos break-through, but this trend is subsequently reversed, causing a gradual decrease in relative concentration values. The experiment C rate constant is nearly four times higher than that for methyl parathion (experiment D). This tap water effect (observed only with chlorpyrifos) is postulated to be due to the presence in tap water of anions that can replace hydroxide and also to possible anion interactions with metallic cations. Eto (*12*) discusses a number of cations, including Cu(II), Ag(I), Ni(II), Co(II), and Zn(II), that catalyze organophosphate, especially phosphorothionate, hydrolysis. Three studies, conducted by Blanchet and St-George (*13*), Meikle and Youngson (*14*), and Mortland and Raman (*15*), detail the catalytic effect of Cu(II) on chlorpyrifos. Cu(II) was shown to be the most effective cation at pH values in the neutral range. Cations present in tap water may bind to resin anions and aid in catalyzing chlorpyrifos on-column hydrolysis. This hydrolysis may match and exceed the relative decrease in hydrolysis due to resin deactivation by anions and exhaustion by chlorpyrifos.

Malathion. Malathion experiment A (Figure 5) is similar in rate to chlorpyrifos and methyl parathion experiments A. Break-through curve shapes are similar to the S-shape observed with methyl parathion. However, experiment B differs from the methyl parathion and chlorpyrifos patterns in that increased flow rate seems to have little effect on the amount of overall degradation observed. Substitution of tap for distilled water at 0.9 mL min^{-1} (C) mirrors the pattern observed for methyl parathion. A rapid exhaustion of the resin followed by malathion break-through occurs; the observed reaction rate constant is nearly five times lower than for distilled water.

The malathion decay product, diethyl fumarate, unlike 3,5,6-trichloro-2-pyridinol and *p*-nitrophenol, is not recovered on the resin. It moves through the resin and into solution, reaching levels of 5.8×10^{-5}, 8.5×10^{-5}, and 1.1×10^{-4} *M* diethyl fumarate in the effluents of Experiments A, B, and C, respectively. Diethyl fumarate does not have the weak organic acid character of trichloropyridinol and nitrophenol, and therefore does not exchange with resin active sites, but exits in the effluent. Malathion decay on the resin occurs so rapidly that, at the flow rates studied, flow rate does not affect degradation. However, replacement of distilled with tap water significantly decreases malathion degradation.

Methamidophos. Methamidophos experiments B (18.5 mg loaded) and C (15.5 mg loaded) demonstrate (Figure 6) that for roughly similar flow rates and amounts loaded, degradation is similar. In addition, the shape of the B and C break-through curves are similar even though concentrations and volumes loaded differ. Thus, it is the overall amount loaded that is important. The break-through curves of experiments A and D, although retaining the sigmoidal shape, differ mainly in their initial relative effluent concentrations. In the case of A, the resin is initially in contact with twice the amount of material at nearly ten times higher flow rate. This translates into a higher initial break-through for experiment A. This set of methamidophos experiments demonstrates the integrated effect of experimental conditions on break-through curve shapes and experimental degradation rates.

Summary. Insecticide hydrolysis by dynamic column RIEX depends on a number of experimental factors, including flow rate across the resin, amount of resin packed, insecticide mass loaded, and the insecticide matrix. The first two variables as well as the third variable (if known), can be successfully controlled to maximize the degradation by RIEX. The fourth variable is not easily controlled in the field, since treatment solutions are likely to be contaminated with foreign materials. The use of tap water in place of distilled water can significantly decrease the resin's degradation capability by factors of 1.5 to 6.5, depending on the actual insecticide. A palpable advantage to RIEX is that for certain insecticides, the decay product can be captured on the resin, if necessary, or discarded and the resin regenerated for reuse.

Conclusion

The organophosphates studied are rapidly degraded by flow-through reactive ion exchange on a resin with hydroxide active sites. This system quickly and efficiently brings reactants into contact, generating degradation rates that are much higher than any other system studied. Comparable resin studies in a static batch system demonstrate much lower reaction rates, with significantly more time needed to achieve similar degrees of degradation. The advantages of reactive ion exchange include stabilization of hydroxide on the resin, adsorption of certain (but not all) decay products to the resin, and efficient reaction in a flow-through system. Under the right conditions, this can produce a purified effluent free of any treatment chemicals. Disadvantages include failure to bind the decay product if it does not exchange with the resin and facile deactivation by contaminating anions. Field mixtures of insecticides commonly contain contaminants such as salts, solvents and soils. These could deactivate the resin, and some filtration would be necessary to prevent column clogging. Thus reactive ion exchange may not be a practical option for intensive field use.

The other two hydrolysis methods examined, aqueous base hydrolysis and perhydroxyl ion hydrolysis, have also been evaluated for their applicability to field use. The first is relatively slow at degrading methyl parathion and requires large amounts of base. The high pH of such treatment solutions would require field remediation by addition of large amounts of acid, and this method would not be convenient for extensive field treatment. The second method, which utilizes sodium perborate, is effective at lower pH values. The addition of large amounts of perborate can significantly lower the pH needed to achieve a given rate of hydrolysis, but the use of a boron derivative may be of environmental concern.

In summary, a number of methods for the hydrolytic degradation of organophosphorus insecticides have been evaluated in this paper. The approach has been to test the methods not only for their efficacy but also with an eye toward their field applicability. We have identified for each method certain advantages and disadvantages. Further research will aid in a final decision of which method is the most practical for the field.

Acknowledgments

David M. Crohn made valuable suggestions about this work. The USDA Regional Research Fund, the Northeast National Pesticide Impact Assessment Program, and the New York State College of Human Ecology-Cornell provided research funding. Hewlett-Packard donated a gas chromatograph/mass selective detector that was invaluable in products identification.

Literature Cited

1. Seiber, J. N. In *Pesticide Waste Disposal Technology*; Bridges, J. S. & Dempsey, C. R., Eds.; Noyes Data: Park Ridge, New Jersey, 1988; pp 187-195.

2. Ketelaar, J. A. A. *Rec. Trav. Chim.* **1950**, *69*, pp 649-658.
3. Lee, G.; Kenley, R. A.; Winterle, J. S. In *Treatment and Disposal of Pesticide Wastes*; Krueger, R. F. and Seiber, J. N., Eds.; ACS Symposium Series 259; American Chemical Society: Washington, DC, 1984; pp 211- 219.
4. Qian, C. F.; Sanders, P. F.; Seiber, J. N. *Bull. Envir. Contam. Toxicol.* **1985**, *35*, pp 682-688.
5. Larsen, L. A. In *Handbook on Toxicity of Inorganic Compounds*; Seiler, H. G. and Sigel, H., Eds.; Marcel Dekker: New York, 1988, pp 129-141.
6. Janauer, G. E.; Costello, M.; Stude, H.; Chan, P.; Zabarnick, S. *Trace Subst. Environ. Health* **1980**, *15*, pp 425-435.
7. Lemley, A. T.; Zhong, W.-Z.; Janauer, G. E.; Rossi, R. In *Treatment & Disposal of Pesticide Wastes*; Krueger, R. F. and Seiber, J. N., Eds.; ACS Symposium Series 259; American Chemical Society: Washington, DC, 1984; pp 245-260.
8. Wolfe, N. L.; Zepp, R. G.; Gordon, J. A.; Baughman, G. L.; Cline, D. M. *Environ. Sci. Tech.* **1977**, *11*, pp 88-93.
9. Fest, C.; Schmidt, K.-J. *The Chemistry of Organophosphorus Pesticides*; Springer-Verlag: Berlin, 1982; pp 25-35.
10. Quistad, G. B.; Fukuto, T. R.; Metcalf, R. L. *J. Agri. Food Chem.* **1970**, *18*, pp 189-194.
11. Lemley, A. T.; Zhong, W.-Z. *J. Envir. Sci. Health* **1983**, *B18*, pp 189-206.
12. Eto, M. *Organophosphorus Pesticides: Organic and Biologic Chemistry*; CRC Press: Cleveland, 1974; pp 78.
13. Blanchet, P.-F.; St-George, A. *Pestic. Sci.* **1982**, *13*, pp 85-91.
14. Meikle, R.W.; Youngson, C.R. *Arch. Environ. Contam. Toxicol.* **1978**, *7*, pp 13-22.
15. Mortland, M.M.; Raman, K.V. *J. Agr. Food Chem.* **1967**, *15*, pp 163-167.

RECEIVED June 1, 1992

Chapter 16

Application Equipment Technology To Protect the Environment

Robert E. Wolf and Loren E. Bode

Department of Agricultural Engineering, University of Illinois, Urbana, IL 61801

The need to protect our environment from the hazards of using pesticides has sparked several technological improvements in application equipment. Efforts to increase operator safety, improve application efficiency and effectiveness, and consideration of ways to reduce amounts of pesticides applied are influencing equipment developments. Researchers are evaluating ways to reduce the drift of pesticides from treated areas. Also, efforts to reduce exposure to those who mix, load, and handle pesticides are being made. Containment structures and mixing-loading pads are being constructed to protect the ground water.

All users of pesticides are confronted with several potential hazards. Those who mix, load, apply, and handle pesticides have a risk of exposure, not only to themselves, but also to the environment. Misapplication, spills, and unsafe application techniques are all major sources for contamination to human, wildlife, and water resources. Since pesticides are likely to be a part of the pest-management system for the foreseeable future, ways to reduce risks caused by pesticides must be practiced (*1*). Five overlapping strategies to help accomplish risk reduction have been proposed (*2*). The strategies are to reduce the amount of pesticide per application, to reduce the number of applications, use of protective clothing and other safety gear, train users in the safe handling and application of pesticides, and change the way the pesticide is applied.

The need to protect our environment from the above hazards has sparked several technological improvements in application equipment. Improvements in both dry and liquid application equipment have been introduced. Direct injection, closed handling systems, on-board dry and liquid application systems, control systems, spot sprayers, shielded sprayers, and tank rinsing devices are examples of technological changes that have affected pesticide contamination to the environment. There has also been a major effort to reduce the amount of chemicals used. Chemical

0097–6156/92/0510–0195$06.00/0

companies are developing new products that are effective at very low rates, but designed for targeted applications with equipment that can precisely apply when and where needed.

Efficient use of inputs has always been the goal of agriculture. Farmers and chemical dealers are becoming more sophisticated and have attitudes that reflect a concern for the environment. Due to public scrutiny of chemical use and regulations limiting the use of agricultural chemicals, it is essential that technological developments are forthcoming to address environmental concerns. Most dealers and growers are ready to evaluate any new developments or practices that are developed. This paper will examine some of the new technology available for pesticide application that will protect the environment from pesticide contamination.

Direct Injection

Direct injection may be the technology that potentially could have the greatest affect on the method of applying pesticides. With direct injection, the spray tank contains only water or carrier. Chemical formulations or specially blended materials are injected directly into the spray lines that are applying the carrier (*3*). The type of mixing that occurs depends on whether the injection occurs before or after the carrier spray pump. Injection systems can be classified by the type of metering pump used. The two systems currently on the market use either a piston metering pump and injects the chemical into the carrier where it is then combined in an in-line mixer prior to spraying or a series of peristalic pumps that meter the chemical and injects it on the inlet side of the carrier spray pump.

The early direct injection systems had several limitations. These included a lag time for the chemical to reach the nozzles, improper mixing of the chemical before spraying, and the units were not adapted for wettable powder formulations (*4*). Many of the early problems with this technology have been resolved. Improved metering pump systems have reduced chemical lag time. In line mixers have resulted in more uniform mixing. The addition of agitation to mix wettable powders, allows the use of a wide variety of formulations. Direct injection technology is available for farm-sized sprayers was well as commercial applicator equipment. Control of injection with computers makes this technology well suited to adjusting rates on-the-go. Rates can be accurately controlled to take advantage of site-specific needs requiring precise application. On line printers are available to produce a permanent record of chemical use and job location.

Direct injection systems are available to simultaneously apply from one to three chemicals at a time. Each chemical requires a separate pump and returnable storage container. The operator can adjust rate and type of chemical with on and off control at any time.

The acceptance of direct injection technology has been spurred by environmental concerns, concern for operator safety, regulations and the development of new products that are effective at very low rates. Direct injection eliminates the need to tank mix chemicals, thus pesticide compatibility problems are eliminated (*5*). Cleanup of equipment is minimal and with no leftover solutions, disposal of rinsates is not a major concern. If the chemicals are in returnable containers and are handled in a closed system, the potential of operator exposure is greatly reduced. Because of the added precision and the ability to spot spray only where the pesticides are needed with the direct injection process, a substantial savings to the producer and the environment is also realized.

On-Board Impregnation Systems

Another technology that has gained widespread acceptance is on-board, on-the-go impregnation of fertilizer and herbicide products. Impregnation, the combination of liquid herbicides and fertilizer for one-pass application, originally accomplished in the fertilizer plant, can now be done with air-flow applicator units that are designed to place herbicide on the fertilizer carrier at the time of the fertilizer application in the field. Introduction of air-flow applicators paved the way for this technique. On-the-go impregnation provides benefits to both the environment and the equipment operators (*6*).

A major environmental improvement with on-board impregnation is moving the impregnation process from the fertilizer facility to the field where the application takes place. Elimination of herbicide residues in the mixing equipment, odors and contaminated dusts at the plant, and reduced operator exposure are all positive factors for on-the-go impregnation. Another consideration is the elimination of potential unused impregnated fertilizer leftover from mixing of excess material.

On-the-go technology is also an advantage to the commercial application businesses because of the opportunities resulting from better and more efficient use of employee time and general reduction in employee exposure to the pesticides being used. Farmers also benefit because of the reduced field compaction with less trips across the field.

In the initial systems, some herbicides were not well suited for the impregnation process. Air machines were not able to properly distribute certain impregnated mixes resulting in clogged distribution systems (*7*). Operator down time to clean and extra maintenance precautions were required to keep the system operating. Experience by the operators and a knowledge of which chemicals work satisfactorily with the on-the-go process have helped this practice remain widely used. Equipment companies continue to make design changes with the air-flow application systems to improve the technique.

With the availability of new granular herbicide formulations, application equipment is being designed to apply dry fertilizer and dry granular herbicides simultaneously. This co-application has become a popular alternative to the original liquid impregnation process (*8*). There are now several granular herbicide products on the market that are capable of being bulk handled in closed systems and can be either co-applied or applied separately. The closed handling systems also protects the operator from unnecessary exposure to the chemical. The co-application process offers many of the same advantages of impregnation while additionally limiting the need to handle liquid chemicals and associated container disposal problems.

The most recent development with liquid impregnation and co-application is the concept of prescription application. In an effort to react to the environmental goals of reduced chemical inputs, equipment manufacturers are developing computer controlled systems that place the fertilizer and pesticides in the exact location and at the precise level needed. This technology is commonly referred to as "variable rate technology" (*9*). Traditional approaches to farming a large field as one unit are becoming obsolete. Since most fields contain different soils and different production capabilities, it is reasonable to apply the inputs based on the variable needs within the field. The term "prescription farming" is often used to describe this practice.

Currently, two systems are being used to apply variable rates of fertilizer and herbicides. For one system, the key to applying the fertilizer and pesticide inputs at variable rates is the development of extensive field maps based on soil tests (*10*). Grid maps for each field are developed with specific soil information provided to help decide the desired input level. The soil map information is also placed on a computer in the applicator vehicle which directs the timing and placement of the fertilizer and herbicide inputs (*11*).

Another available system uses a soil probe mounted on the front of the applicator unit to analyze the soil organic matter on-the-go, transmits the information through a computer, and regulates the amount of fertilizer applied to the field (*12*). This process does not involve extensive preliminary field testing. However, to be most effective, the system works best with soils which exhibit differences in organic matter. It is also dependent on the development of sensors to detect soil organic matter.

Handling Systems

A major emphasis by chemical companies and equipment manufacturers has been to develop new and innovative ways to make the handling of chemicals more convenient and to reduce exposure for the people who use pesticide products. Bulk and mini-bulk handling systems are available to store, transport, and handle liquid and granular pesticides. The closed systems associated with bulk tanks reduce operator contact with the chemicals, eliminate potential spillage, and with the returnable 250 to 300 gallon containers, container disposal is eliminated. Commercial and private applicators can now purchase and use pesticide products with reduced exposure and the returnable containers eliminate the disposal problems associated with nonreturnable containers.

Closed handling systems are also being developed to store, transport, and transfer dry granular insecticides (*13*). Pneumatic handling of granular products is being used to transfer granular herbicides from bulk storage at the plant to the on-the-go applicator units in the commercial industry.

Control Systems

The driving force behind much of the previously discussed application technology is the development of sensors and application of controllers. Spray controllers are being integrated in spray monitor systems. Electronic devices to control application rates have been widely used for years (*14*). Controllers are designed to automatically compensate for changes in speed and application rates on-the-go. Some are computer-based and work well with new application techniques such as direct injection and variable rate application. Computers and controllers work together to place fertilizer and pesticide inputs in the precise position at the prescribed amount.

Tank Rinsing Systems

A major environmental problem that applicators have faced is handling left over chemical mixes. It is impossible to completely empty a spray tank. Miscalculations of amounts of spray material needed also lead to leftover mixes that require disposal. Another concern is how to handle materials used to rinse out and clean a sprayer at the end of a job or the season. Several alternatives are available, but two are the most practical at this time. In some states regulations dictate that a rinsate pad and specific rinsate storage containers must be used to collect and recycle the rinse materials (*15*). Another alternative involves the inclusion of special tank rinsing nozzles that can be incorporated into the spray system. Tank rinsing nozzles are available for both commercial and farm sprayers. For these nozzles to be effective, a fresh water tank is mounted on the sprayer. The fresh water is used to rinse the outside of the spray equipment and with the special nozzle arrangement can rinse and clean the inside of the spray tank.

The rinsate pad provides the applicator a place to clean and rinse the sprayer. Storage of the cleaning solution and tank rinsate for later field application is an acceptable alternative for properly disposing of the rinsate materials. Field rinsing allows the operator to avoid cleaning the equipment at the mixing and loading site where potential environmental contamination can occur.

Shielded Sprayers

Technological developments to reduce the amount of off-target drift are of major interest. Equipment companies are using hoods, cones, or similar devices to protect the spray from winds that move spray droplets off-target (*16*). Other companies are working with shielded booms for reducing drift. Shielded booms use an downward air current to direct the spray droplets to the target. These systems are still being developed and evaluated to determine their effectiveness.

Future Developments

Technological improvements in the application industry have occurred at very rapid rate in recent years. As scientists continue to focus in on the precise farming of tomorrow, the equipment industry will work to improve and develop equipment needed to achieve the goal of more effective application. Major developments in field mapping and computer application controls are being refined. Use of satellites as a method of controlling field positioning is being developed. Accurate field positioning and extensive field mapping for chemicals will provide a precise system for meeting the application needs of the future.

Literature Cited

(*1*) Bode, L.E. *American Journal of Industrial Medicine* **1990**, *18*, 485-9.
(*2*) Dover, M.J. Holmes, PA, 1985, World Resources Institute, Study 4, 84.
(*3*) Finck, C. *Farm Journal* **1991**, *115* (8), 14-15.
(*4*) Bode, L.E.; Wolf, R.E. "Proceedings of Papers" 43rd Annual Illinois Agricultural Pesticides Conference, Jan. 1991, University of Illinois: Urbana, Il, 1991; 170-1.
(*5*) Pocock, J. *Prairie Farmer* **1991**, *163* (1), 12-4.
(*6*) ______. *Agrichemical Age* **1989**, *33* (2), 6-7.
(*7*) ______. *Farm Industry News* **1989**, *22* (1), 93.
(*8*) Klassen, P. *Farm Chemicals* 1990, *153* (2), 63-5.
(*9*) Reichenberger, L.; Russnogle, J. *Farm Journal* **1989**, *113* (6), 11-3.
(*10*) Simmonds, B.; Brosten, D. *Agrichemical Age* **1991**, *34* (2), 6-9.
(*11*) Brunoehler, R. *Farm Industry News* **1991**, *24* (12), 22-3.
(*12*) ______. Farm Chemicals **1990**, *154* (11), 36-7.
(*13*) Landphair, D.K.; Olson, G.M.; Tenne, F.D. American Society of Agricultural Engineers, New Orleans, LA, Dec. 1989; ASAE: St. Joseph, MI; Paper #891639.
(*14*) Finck, C. *Successful Farming* **1990**, *88* (3), 14-5.
(*15*) Kammel, D.W.; Noyes, R.T.; Riskowski, G.L.; Hofman, V.L. "Designing Facilities for Pesticide and Fertilizer Containment"; Midwest Plan Service: Ames, Iowa, 1991; 3-4.
(*16*) Maybank, J.; Shewchuk, S.R.; Wallace, K. *Canadian Agricultural Engineering* **1990**, E-906-16-A-89, 235-41.

RECEIVED March 30, 1992

Chapter 17

Photochemical and Microbial Degradation Technologies To Remove Toxic Chemicals

Fumio Matsumura and Arata Katayama

Department of Environmental Toxicology, University of California, Davis, CA 95616

An effort was made to apply photochemical degradation technology on biodegradation processes to increase the bioremediation potential of microbial actions. For this purpose, we have chosen *Phanerochaete chrysosporium*, a wood decaying white-rot fungus and a variety of chlorinated pesticides and aromatics as study materials. By using UV-irradiation and benomyl (a commonly used fungicide) as selection methods, a strain of UV-resistant *P. chrysosporium* was developed. This strain was found to be capable of rapidly degrading these chlorinated chemicals when they were incubated in N-deficient medium which received 1 hr/day of UV-irradiation. UV-irradiation either at 300 or 254 nm showed the beneficial effect of speeding up the rate of degradation on most of test chemicals with the exception of toxaphene and HCH (hexachlorocyclohexane). By adding fresh glucose to the medium it was possible to maintain high degradation capacity for several weeks.

The presence of highly toxic chemical residues in wastes creates rather difficult problems for their containment, transportation and disposal. For example, first, such toxic chemicals could render the entire waste "hazardous" requiring very careful handling; second, safety questions become critical issues even when the level of contamination is very low; and third, their presence makes it difficult to plan remediation strategies since selective elimination of small amounts of toxic material is a tremendous challenge. In addition, some toxic chemicals are very stable in the environment, resisting general environmental degradative forces.

From the viewpoint of microbial degradation strategies, there are additional problems with these chemicals. They are present in rather low quantities, making it difficult to develop specific microorganisms to utilize them as sole carbon-energy sources. Their accessibilities are generally poor, because many of them are bound tightly to soil particles. And many of these chemicals contain halogenated moieties which are foreign to the microbial world, and, therefore, are not generally attacked by microbial enzymes.

In view of these difficulties, we have been experimenting with the idea of combining microbial technologies with photochemical degradation technologies.

0097–6156/92/0510–0201$06.00/0

Photochemical technologies offer unique advantages over other approaches to chemical degradation, summarized as follows: First, they are easy to manipulate and standardize, being physical forces; second, they are nonselective and largely concentration-independent, being capable of degrading chemicals at low concentrations as well as in the presence of other chemicals; and third, they are powerful enough to attack carbon-halogen bonds to alter chemicals to less halogenated forms which in turn could be degraded by biological forces. Also, there are several methods to enhance photochemical reactions such as addition of ozone, hydrogen donors, photosensitizers and nucleophilic and/or electrophilic agents.

On the other hand, photochemical technologies have one notable shortcoming. That is, the path of light can be readily blocked by solid objects or light absorbing substances. Thus, toxic chemicals buried under soil or solid wastes cannot be degraded by photochemical means, and those present in turbid, murky solutions cannot readily be degraded.

Nevertheless, there are many potential advantages of combining these two major technologies, since they are based on two totally different principles, and since both approaches have been extensively studied and have reached the stage where they could be considered as viable approaches to *in situ* remediation of chemicals.

Biodegradation of persistent halogenated organic chemicals such as PCBs, dioxins, chlorinated pesticides etc., are of great interest from the viewpoint of their potential use to clean up the contaminated sites and industrial waste streams on-site (i.e., *in situ* remediation) (*1-3*). Many of these compounds are known to be highly toxic (*4*), and furthermore, other methods such as incineration are not only expensive but also known to cause secondary pollution problems (*5*). Recent studies have shown that lignin-degrading white rot fungi possess capabilities to degrade a variety of highly recalcitrant and toxic compounds (*6-10*). These are organisms which require oxygen for growth. Therefore, they spread on the surface of substrates (e.g., soil, trees, etc.) and thereby expose themselves to sunlight. Thus, it occurred to us that these fungi may be good candidates to test the feasibility of combining microbial with photochemical degradation technologies (*11*).

To be sure, there have been several attempts in the recent past to combine these two technologies. For example, it has been observed that irradiation by simulated sunlight increased the mineralization rate of 4-chlorobiphenyl in river sediment containing a mixed microbial population (*12*), and that microbial actions by a *Pseudomonas* sp. followed by subsequent irradiation by simulated sunlight degraded the yellow metabolites of 2,4-dichlorobiphenyl (i.e., sequential treatment) (*13*). In another study, Kearney et al. (*14*) first treated [^{14}C]-2,4-6-trinitrotoluene (TNT) by ultraviolet ozonation and then subjected the products to microbial degradation by *Pseudomonas putida*. They found that the former treatment helped the metabolic degradation of TNT. However, there has been no successful demonstration of simultaneous application of these two technologies (i.e., use pure cultures of microbial and ultraviolet treatments for the degradation of highly recalcitrant compounds). The main obstacle in developing such simultaneous combination systems has been the susceptibilities of microorganisms in general to UV irradiation. To overcome this problem, we have developed an ultraviolet- and fungicide-resistant stain of white rot fungus. The following is a brief description of our basic approach and preliminary results. The details and complete results of this particular study on *Phanerochaete chrysosporium* will be published elsewhere as a technical publication.

Materials and Methods

Selection of Microbial Strains. For screening of fungi for UV resistance, a piece of hypha was placed on the center of a 3% malt extract agar plate and incubated in the dark at 25°C to allow the linear elongation stage of hyphae. The culture was irradiated through a polystyrene Petri dish cover (90% UV cutoff at 290 nm) with 7000 μW/cm^2 of UV at 300 nm for 2 h/day during regular incubation at 25°C. The hyphae growth was measured as the increase in diameter of the colony and compared with the growth rate obtained under an identical but nonirradiating condition. White rot fungi tested in addition to *Phanerochaete chrysosporium* were three strains of *Armillaria mellea*, two strains of *Ganoderma lucidum*, a strain each of *Ganoderma, brownii*, *Inonotus cuticularis*, *Lentinus betulia*, *Phellinus gilvus*, and *Trametes hirsuta*, and two strains of *Trametes, versicolor*. From these screening processes, one strain of *P. chrysosporium*, BKM F-1767, obtained from Dr. T.K. Kirk's group has emerged as the most UV resistant isolate by the criteria shown later in this paper. It was further subjected to selection for both benomyl and UV tolerance as follows: conidia were incubated on YM agar plate containing 100 mg/L benomyl at 25°C. The colony grown were transferred to nitrogen-deficient broth (*15*) and routinely irradiated with UV at 300 nm 2 h/day for 4 weeks. Since no sporulation occurred under these conditions, a piece of hypha was transferred to YM agar slant to obtain conidia. The process was repeated several times. During such processes, several isolates were obtained showing the higher ability to degrade 3,4,3',4'-tetrachlorobiphenyl (TCB) in the fungus-UV combination system ("Petri dish method", see below) than in the original strain BKM F-1767 alone. The most resistant strain was designed as the BU-1 strain.

Studies on Metabolism of [^{14}-C]TCDD. The nutrient solution was the same as that reported by Kirk et al. (*15*), except for the use of 10 nM, 2,2-dimethylsuccinate buffer (pH = 4.5). For this purpose, 20-mL scintillation glass vials (90% of UV cutoff at 280 nm), each equipped with a cap with an inlet and an outlet needle, were used. About 7 x 10^5 conidia were inoculated into 1 mL of the nutrient solution in each vial and incubated for 3 days at 25°C to allow formation of the mycelial mat. To each vial was added 0.649 μg of 2,3,7,8-tetrachlorodibenzo-*p*-dioxin (TCDD) [^{14}C uniformly ring labeled, 365 μCi/mg custom synthesized by Amersham, purity > 99% by TLC (*16*)] with 35 μL of *p*-dioxane. UV irradiation was carried out at 300 nm, 3400 μW/cm^2 for 2 h every day, using a 15-W midrange UV bulb purchased from Fotodyne, New Berlin, WI. Flasks were flushed every day with room air. The produced $^{14}CO_2$ and the volatilized intermediates or parent compound were trapped by the method of Marinucci and Bartha (*17*).

Degradation Studies Using the Petri Dish Method. The standard method employed in studying the disappearance of the originally added halogenated pollutants was the Petri dish method. For this purpose, cultures of *P. chrysosporium* in nitrogen-deficient medium (2 mL) as described above were incubated in 35-mm-diameter polystyrene Petri dishes. After 3 days of preincubation at 27°C in the dark, each pollutant was added (1 µg/mL) and the culture was irradiated with 300- or 254-nm light for 2 h/day. Polystyrene covers (90% of UV cutoff at 290 nm) were used for 300-nm UV (3400 $\mu W/cm^2$) irradiation, and polyvinyllidine covers (Saran Wrap, 90% of UV cutoff at 220 nm) were used for 254-nm UV (800 $\mu W/cm^2$) irradiation. After incubation, the culture was extracted by use of a hand homogenizer (Wheaton) with 0.4 mL of methanol. A 3-mL mixture of hexane (bp 68-70°C) and 40% $CaCl_2$ (2:1) was added, vortexed thoroughly, and centrifuged at 700 g for 1 min. For TCB, Arochlor 1254, and TCDD, benzene was used in place of hexanes. Routinely, a fraction of the organic solvent layer was directly analyzed by gas chromatography (GLC) because no serious interference peaks coinciding with the analytes were observed. In the case of Arochlor 1254 only, the extract was washed with concentrated H_2SO_4, 5% Na_2CO_3, and 20% NaCl solution (*18*) prior to GLC analyses. The instruments used were a Varian gas chromatograph (Model 2400) equipped with 1.5% OV-17/1.95% OV-210 on Chromosorb AW-DMCS 80/100 mesh for other pollutants, and a ^{63}Ni electron capture detector coupled to a Waters Maxima 820 chromatography workstation for quantitative analysis for peak integration. Detector and injector temperatures were 265 and 245°C, and the column temperatures were 235, 225, 225, 215, 225, 215, and 200°C for DDT, dieldrin, heptachlor, toxaphene, TCB, Aroclor 1254, and TCDD, respectively. Only in the case of [^{14}C]TCDD, the reaction product was extracted as above and further analyzed by a silica gel thin-layer chromatography method using CCl_4 as a mobile phase (*16*) and the remaining [^{14}C]TCDD was counted by liquid scintillation. The detection limit was 6.7 pg/ml TCDD.

Results and Discussion

Several white rot fungi were screened for their native sensitivities toward UV irradiation. Of these, a strain of *P. chrysosporium*, BKM F-1767, was found to be most resistant to both UV and benomyl. The fungal elongation rate on malt agar irradiated by UV at 300 nm for 2 h/day was approximately 80% of that in the nonirradiated condition. From this original strain a UV-selected and benomyl-resistant *P. chrysosporium* BU-1 was selected for further testing. This strain was capable of sustaining its growth when subjected to UV irradiation 2 h/day for indefinite time periods if proper carbon and nutrient supplements were provided.

The initial study results using [U-^{14}C]TCDD established conclusively that simultaneous application of UV irradiation (at 300 nm, 2 h/day) and this strain of *P. chrysosporium* in a nitrogen-deficient medium caused a much more accelerated rate of mineralization of this compound than those achieved by either photochemical action alone or microbial action alone. The mineralization continued at a steady rate until the end of the experiment, and the amount of $^{14}CO_2$ released reached 20% of the initial radioactivity after 40 days of incubation. In the case of "fungus only" tests the corresponding figure was 5%, and the "blank" (i.e., no UV, no fungus) figure was less than 0.1%. The recovery of radioactivity measured after 7 days of incubation in the flask was found to be 100.0 ± 1.6%. Volatilized total radioactivities (i.e., solvent-extractable metabolic intermediates or parent compound trapped in a glass wool plug) detected after 40 days of incubation were as follows: 0.39% of the initial radioactivity in the combination system, 0.15% in the UV-only system, and 0.015% in the non-fungus and non-UV systems, indicating that evaporation of the original compound and/or its degradation product plays only a minor role in the overall fate of TCDD in this reaction system. Together these results clearly showed that UV irradiation could act synergistically with the microbial degradation activities. In parallel experiments TCDD was treated in the same medium by UV irradiation only, resulting in a mineralization level of 5.8%. The corresponding figures were 0.27% in the test with the fungus only and 0.19% without irradiation and without inoculation of fungus in the same medium.

In the second series of experiments, the rates of disappearance of DDT, dieldrin, heptachlor, 3,4,3'4'-tetrachlorobiphenyl, toxaphene, and TCDD were studied by using the Petri dish method, where the disappearance of the original compounds were monitored by gas chromatography (Figure 1). By the simultaneous treatments of the fungus and UV either at 254 or 300 nm, more than 97% of the initial added amounts of DDT, TCDD, heptachlor, and TCB were metabolized in 3 weeks of incubation. By combination with UV at 254 nm, 90 ± 1% of dieldrin and 52 ± 1% of toxaphene was degraded in 4 weeks. In the same manner, some additional halogenated chemicals were also tested. Those readily degraded by this system were endosulfan (57 ± 1% degradation after 2 days by combination with UV at 254 nm), pentachlorophenol (98 ± 1% after 28 days with UV at 254 nm), and that on which the system worked slowly was γ-BHC (26 ± 11% after 28 days with UV at 300 nm). In view of the past criticism in this field that microbial degradation technologies tend to have a common problem in not eliminating the last few fractions of the toxics, we have made a conscious effort to examine whether degradation by this system could continue to the level where we could no longer find their residues by the available analytical techniques. In the data shown in Figure 1, we could show that in four of the fast degrading cases, the residue levels reached the GLC detection limits for each compound within the study period (1 ng/mL DDT, 29 ng/mL TCDD, 6 ng/mL TCB, and 8 ng/mL heptachlor). Furthermore, study by Petri dish method using [U-^{14}C]TCDD showed that 8.8 ng/mL initial concentration could be reduced to 31 ± 10 pg/mL after 7 days of incubation and to the approximate ^{14}C-radioassay detection limit of 8.0 ± 2.2 pg/mL after 28 days.

By use of DDT, toxaphene, TCB, and TCDD, the synergistic and antagonistic effects of the combination treatment on the rates of degradation were examined (Table I). In the case of degradation of TCB and TCDD, the combination of these two technologies always produced synergistic actions. In the degradation of DDT and toxaphene, the action of UV at 300 nm was not synergistic with microbial actions, though UV at 254-nm irradiation always caused higher rates of degradation than those obtained by the action of the fungus alone.

The degrading ability of *P. chrysosporium* itself is generally considered to be largely due to the lignin-degrading enzyme system produced in nitrogen-deficient culture (*6*). In the combination test with the fungus and UV at 300 nm, we found that the degradation rate of TCB (10 mg/L initial concentration) was higher in nitrogen-deficient culture than in nitrogen-rich culture, which was prepared by the addition of 1 mg of NH_4NO_3. On the other hand, the autoclaved culture containing dead fungal mycelia and spent medium also enhanced the degradative action of UV to a significant extent. These findings suggest that the lignin-degrading enzyme system

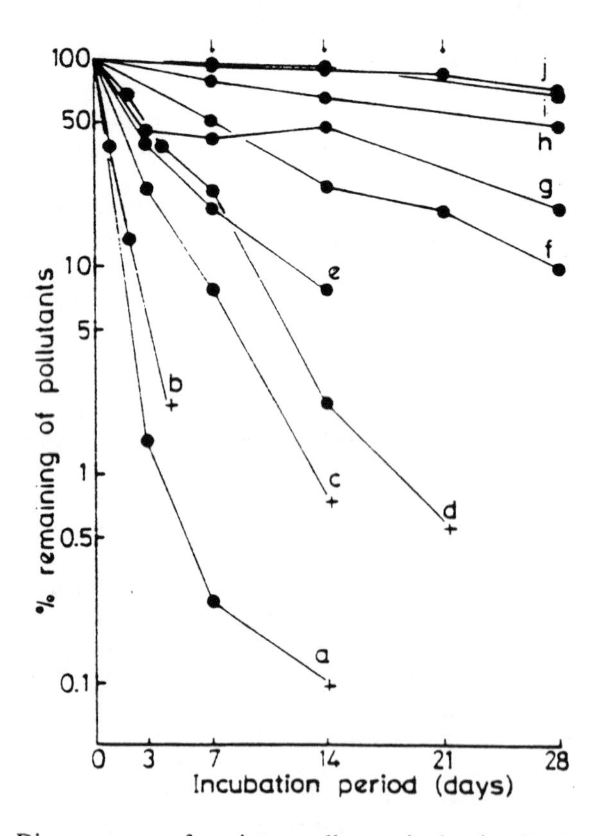

Figure 1. Disappearance of persistent pollutants in the simultaneous treatment using *P. chrysosporium* BU-1 and UV (either 254 or 300 nm) irradiation. (a) DDT in combination with 254-nm UV, (b) TCDD with 300-nm UV, (c) heptachlor with 254-nm UV, (d) TCB with 300-nm UV, (e) heptachlor with 300-nm UV, (f) dieldrin with 254-nm UV, (g) DDT 300-nm UV, (h) toxaphene with 254-nm UV, (i) toxaphene with 300-nm UV, (j) dieldrin with 300-nm UV. The values are duplicate flask averages of two independent tests and are expressed as percents of the initial amount (1 mg) of pollutants. Duplicate tests varied less than ± 10%. The sign + indicates that the residue level reached less than the detection limits for each compound under the analytical technique used. Glucose (10 mg) was added every 7 days (shown as arrows). (Reproduced from reference 22. Copyright 1991 American Chemical Society.)

is an important factor, but not the sole factor responsible for degradation, and that some microbially produced heat-stable photosensitizers also promote photodegradation of certain chemicals.

It is important to mention here that a glucose supplement is required after certain time intervals to maintain the degrading ability of the combined system. Also, as long as glucose was added periodically, there appears to be no sign of slowing of the degradative activity of this combined system over the experimental period, the longest being 40 days.

Table I. Effect of Short-Wavelength and Long-Wavelength UV Irradiation on the Degradation Activities of *P. chrysosporium* BU-1 on Persistent Pollutants[a]

Pollutants	*Incubation period, days*	*% Remaining after UV irradiation[b]*		
		None	*254 nm*	*300 nm*
DDT	7	24 ± 2	0.2 ± 0.1	42 ± 6
toxaphene	7	91 ± 6	79 ± 3	93 ± 2
TCB	7	73 ± 7	47 ± 10	24 ± 8
TCDD	4	44 ± 19	41 ± 14	3 ± 3

a The effects of UV irradiation only could not be evaluated by petri dish method because of the volatilization loss of the test compounds.
b In the presence of fungus.

At this stage, one could question why a combined technology should offer advantages over each single application technology. The most important considerations in any biodegradation should be the efficiency and the reproducibility of the technology. To this end, photochemical approaches provide, first, superior reproducibility profiles, light being a physical rather than biological force and second, they provide higher energies than are attainable by biological systems alone. Thus, when applied to halogenated aromatics, for instance, UV-irradiation in the presence of hydrogen donors is capable of dislodging halogens directly from aromatic rings. Thus even biologically very stable chlorinated dioxins are readily degraded by UV-irradiation *via* initial dechlorination. Once the number of chlorine atoms per molecule of polychlorinated dioxins are reduced, they are readily attacked by microorganisms that are capable of metabolizing aromatic compounds *via* mono- and dioxygenases. By the same token, UV-irradiation may not offer great advantages in degrading UV-resistant chemicals such as toxaphene and HCH (BHC). Indeed, these chemicals are manufactured by UV-irradiation of starting cyclic hydrocarbon precursors in the presence of Cl_2. In such cases, a better approach would be to dechlorinate these chlorinated alkanes biologically or chemically under anaerobic conditions and then react with oxidative biological or photochemical reactions.

Another important point this combination of technologies offers would be the high feasibility of practical application at larger scales. The reasons for this are: (a) *P. chrysosporium* has already been used by the wood-pulp industry in large scale treatment facilities (*19*); (b) it is known to compete well against other microorganisms under nonsterile conditions (*7, 17, 19*); (c) as shown in this work, it is possible to develop fungicide-resistant strains and, thereby, give *P. chrysosporium* a selective advantage; and (d) we already know that not too many competing organisms are UV resistant, as shown in the current work.

Furthermore, *P. chrysosporium* is capable of growing fast to form mycelial layers (sometimes referred as "mycelial mats") on various surfaces of artificial substrates, particularly those exposed to air and the light. These layers are capable of bioconcentrating lipophilic chemicals such as pesticides and PCBs in aqueous media. Thus, these characteristics help in transporting many of the toxic chemicals to surfaces which UV-light can reach.

Another key point of this system is that, apparently, it works on a very wide spectrum of organic chemicals, probably reflecting the nonspecificity of the lignase because of its nature to bind first with the peroxide molecule to create a reactive complex which, in turn, attacks organochemicals (*10, 15*).

There are two major goals for any biodegradation technologies involving highly toxic chemicals. One is to remove the bulk of toxic compounds and their toxic metabolic products as fast as possible. The other is to eventually reduce their levels to below the critical risk level (*20*).

Certainly no single method will offer panacea to all types of waste or pollution cases. Much more effort must be made to apply this combined technology to individual cases. Furthermore, limitations of this approach as well as specific requirements may have to be defined in the future to make it applicable to situations best suited to such an approach. Nevertheless, what we would like to stress is that this noble idea is worth pursuing. After all, photochemical and microbial degradative reactions are two major types of forces removing toxic chemicals from the natural environment (*21*). For this reason, both simultaneous and sequential application methodologies must be considered important in testing the best approaches for individual chemical and waste disposal cases. Moreover, in the future, utilization of some indigenous species having natural UV-resistance characteristics as well as competitive advantages over introduced species might be feasible, and more economical and efficient. Therefore, the most important message we wish to convey here is the idea of utilizing two entirely different degradation technologies in concert to remove unwanted chemicals accumulating in our environment.

Acknowledgments

We thank Drs. T.K. Kirk, J.E. Adaskaveg, and W.W. Wilcox for supplying the original BKM F-1767 strain of *Phanerochaete chrysosporium*. [U-^{14}C]TCDD was supplied by the Dept. of Environmental Toxicology, College of Agricultural and Environmental Sciences, University of California, Davis and ES04699, Superfund Program grant, from the National Institute of Environmental Health Sciences, Research Triangle Park, NC.

Literature Cited

(*1*) Thomas, J.M.; Ward, C.H. *Environ. Sci. Technol.* **1989**, *23*, 760-766.
(*2*) Daley, P.S. *Environ. Sci. Technol.* **1989**, *23*, 912-916.
(*3*) Alexander, M. *Science* **1981**, *211*, 132-138.
(*4*) Matsumura, F., Ed. *Differential Toxicities of Insecticides and Halogenated Aromatics*; Pergamon Press, Oxford, UK, **1984**; pp 1-588.
(*5*) Rappe, C. *Environ. Sci. Technol.* **1984**, 18, 78A-90A.
(*6*) Bumpus, J.A.; Tien, M.; Wright, D.; Aust, S.D. *Science* **1985**, *228*, 1434-1436.
(*7*) Eaton, D.C. *Enzyme Microb. Technol.* **1985**, *7*, 194-196.
(*8*) Mileski, G.J.; Bumpus, J.A.; Jurek, M.A.; Aust, S.D. *Appl. Environ. Microbiol.* **1988**, *54*, 2885-2889.
(*9*) Hammel, K.E.; Kalyanaraman, B.; Kirk, T.K. *J. Biol. Chem.* **1986**, *261*, 16948-16952.
(*10*) Kirk, T.K.; Farrell, R.L. *Annu. Rev. Microbiol.* **1987**, *41*, 465-505.
(*11*) Zabik, M.J.; Leavitt, R.A.; Su, G.C.C. *Annu. Rev. Entomol.* **1976**, *21*, 61-79.
(*12*) Kong, H.-L.; Sayler, G.S. *Appl. Environ. Microbiol.* **1983**, *46*, 666-672.
(*13*) Baxter, R.M.; Sutherland, D.A. *Environ. Sci. Technol.* **1984**, *18*, 608-610.
(*14*) Kearney, P.C.; Zeng, Q.; Ruth, J.M. *Chemosphere* **1983**, *12*, 1583-1597.
(*15*) Kirk, T.K.; Schultz, E.; Connors, W.V.; Lorentz, L.F.; Zeikus, V.G. *Arch. Microbiol.* **1978**, *117*, 277-285.
(*16*) Quensen, J.F., III; Matsumura, F. *Environ. Toxicol. Chem.* **1983**, *2*, 261-268.
(*17*) Marinucci, A.C.; Bartha, R. *Appl. Environ. Microbiol.* **1979**, *38*, 1020-1022.
(*18*) Murphy, P.G. *Assoc. Off. Anal. Chem.* **1972**, *55*, 1360-1362.
(*19*) Eaton, D.C.; Chang, H.M.; Joyce, T.W.; Jeffries, T.W.; Kirk, T.K. *Tappi* **1982**, *65*, 89-92.
(*20*) Abelson, P.H. *Science* **1989**, *246*, 1097.
(*21*) Matsumura, F. *Ecotoxicology and Climate.* Bourdeau, P.; Haines, J.P.; Klein, W.; Krishna Murti, C.R. Eds.; SCOPE, John Wiley & Sons, New York, **1989**; pp 79-89.
(*22*) Katayama, A.; Matsumura, F. *Environ. Sci. Technol.* **1991,** *25*, 1329–1331.

RECEIVED May 22, 1992

Chapter 18

Biodegradability of Pesticides Sorbed to Activated Carbon

J. H. Massey, T. L. Lavy, and M. A. Fitzgerald

Department of Agronomy, Altheimer Laboratory, University of Arkansas, Fayetteville, AR 72703

Activated carbon filtration of wastewater results in the production of pesticide-containing carbon which requires further disposal. Studies were conducted to assess the biodegradability of carbon-bound pesticides as a potential disposal method for spent carbon. Evolution of CO_2 was not affected by the presence of 10 pesticides bound to actual spent carbon but was inhibited by freshly added 2,4-D and alachlor. After 108 days incubation at 20°C, <0.5% of initially added ^{14}C-2,4-D was recovered as $^{14}CO_2$ from carbon treated with 380 ppm 2,4-D. The age of the carbon-bound pesticides is proposed as an important mechanism controlling the bioavailability of the sorbed pesticides.

Activated carbon adsorption effectively removes many pesticides from leftover pesticide solutions and rinsates (*1,2*). This technology is simple and effective when used to dispose of liquid pesticide wastes. However, a major drawback associated with activated carbon filtration is that spent carbon generated in the wastewater treatment process requires disposal.

Spent carbon is commonly regenerated thermally, but it is not economically feasible when less than 225 kg spent carbon per day are generated (*3*). For this reason, spent carbon generated in small quantities, as under typical pesticide application conditions, is normally incinerated.

Incineration of pesticide-containing wastes is a highly effective means of disposal (*4*), but it does not allow for the on-site disposal of wastes. This is contrary to the recommendations developed in several pesticide waste disposal conferences where it was suggested that pesticide rinsates be treated and disposed of on-site (*5*). The goal of the preliminary studies described in this paper was to assess the feasibility of biologically degrading pesticides bound to carbon as a means of on-site spent carbon disposal.

0097–6156/92/0510–0210$06.00/0

MATERIALS AND METHODS

Effect of Spent Carbon on Microbial Activity. Activity was measured by quantifying the amount of CO_2 evolved from carbon containing pesticides and comparing this to the amount evolved by non-treated carbon (Darco Cullar-D granular activated charcoal, American Norit Co., Inc., Jacksonville, FL 32205) and liquid controls. For this study, ten 60 g amounts of actual spent carbon (Cullar-D charcoal containing acephate, atrazine, carbaryl, carbofuran, metolachlor, fluazifop-butyl, paraquat, pendimethalin, sethoxydim, and trifluralin) was added to 125 ml Erlenmeyer flasks along with 50 ml of nutrient broth solution (4 g/L; BBL Nutrient Broth, Becton Dickinson Microbiology Systems, Cockeysville, MD 21030). The total amount of pesticide active ingredient held by the spent carbon was about 5% (w/w). Sample controls included ten 60 g amounts of untreated carbon to which 50 ml nutrient solution was added, and 10 non-carbon controls consisting of 50 ml nutrient solution. All samples were inoculated with one ml of a 10^{-6} soil dilution and incubated at 24 ± 1° C. The experimental design was a randomized complete block with two replications per treatment and 5 repetitions per replication. Carbon dioxide evolution from spent carbon, untreated carbon, and liquid controls was measured using a flow-through CO_2 trapping apparatus (*6*).

At weekly intervals, two grams of carbon were removed from each of the three replications of spent and new carbon samples. One gram was used to determine the moisture content of each sample by drying at 110° C for 24 hours. The other was added to a 1 cm x 10 cm glass culture tube along with 5 ml of a 0.023 M methylene blue dye solution buffered with 0.108 M sodium phosphate monobasic (pH 6.5). Next, the culture tubes were placed on a rotary shaker for 15 min at 12 rpm. By measuring the absorbance of the dye solution at 664 nm and comparing to a standard curve, the amount of dye adsorbed by the carbon was determined. These adsorption data were statistically analyzed as a split plot with charcoal type as the main-plot and incubation time as the sub-plot.

Degradation of 2,4-D Herbicide Sorbed to Activated Carbon. The degradation of 2,4-D adsorbed to activated carbon was studied by adding 0.36 μCi of 2,4-dichlorophenoxyacetic acid-ring-UL-^{14}C (>98% purity, 12.8 mCi/mmol; Sigma Chemical Co., St. Louis, MO 63178) to either 50 g carbon (Darco Cullar-D granular activated charcoal), 50 g silt loam soil, or 50 ml deionized water. Prior to addition to the samples, the ^{14}C-2,4-D was thoroughly mixed with formulated 2,4-D dimethylamine (47.4%, Red Panther Chemical Co., Clarkdale, MS.) for a final concentration of 380 ppm. To the carbon and soil samples was added 13 ml of a buffered (pH 6.5) nutrient solution (8 g/L nutrient broth + 15 g/L NaH_2PO_4). An equivalent amount of nutrients was added to the liquid controls.

The effect of alachlor on the degradation of 2,4-D adsorbed to carbon was studied by adding 5000 μg alachlor (Lasso 4 EC, Monsanto Agricultural Products Co., St. Louis, MO 63167) to 50 g carbon in addition to 380 ppm 2,4-D as with the other samples. All samples were inoculated with one ml of a 10^{-3} soil dilution, and incubated at 19 ± 2° C. Moisture levels and pH were adjusted weekly to 100% moisture-holding capacity and values near neutrality, respectively.

Evolved $^{14}CO_2$ and CO_2 were trapped using a flow-through apparatus (*6*). Radioactivity was measured by adding 1 ml of trapping KOH to 10 ml scintillation cocktail (ScintiVerse BD, Fisher Scientific, Fair Lawn, NJ 07410) and counting with a Packard Tri-Carb 4530 (Downers Grove, IL 60515) scintillation counter.

RESULTS

Effect of Spent Carbon on Microbial Activity. The cumulative amounts of carbon evolved as CO_2 from pesticide-treated carbon, untreated carbon, and liquid controls are given in Table I. No significant differences (α = 0.05) in CO_2 evolution were observed.

Table I. C-CO_2 Evolved from Spent Carbon After 53 Days Incubation at 24 ± 1°C

Matrix	mg C-CO_2
Liquid	1552
Spent Carbon	1509
New Carbon	1466
LSD(5%)	NS

Table II shows that the amount of dye adsorbed by the spent carbon did not vary over time. The amount of methylene blue dye adsorbed by the carbon served as an indicator for the adsorption capacity of the carbon. The amounts of dye adsorbed by the untreated and spent carbon were significantly different (α = 0.05), reflecting differences in the number of available adsorption sites on the untreated and spent carbon.

Generally, the amount of dye adsorbed over time did not vary (α = 0.05). This suggests that a measurable number of adsorption sites occupied by pesticides on the spent carbon did not become available for dye adsorption over time, i.e. the pesticides were not measurably desorbed or degraded. These results agree with those of Benedek (*7*) who, using iodine adsorption as a measure of carbon adsorption capacity, found little biological regeneration of carbon containing phenol.

It is possible that the pesticides were degraded into products which were also strongly adsorbed by the carbon. In this case, as in the case of limited pesticide desorption and degradation, no increase in dye adsorption would have been observed. Our study could not, however, differentiate between these possibilities.

Table II. Amounts of Methylene Blue Dye Adsorbed by Spent Carbon as a Function of Incubation Time

Incubation Time (d)	mg dye per g Carbon[1]	
	New	Spent
0	30.7	16.6
7	34.7	21.7
17	30.8	17.3
21	31.6	19.9
27	29.9	17.5
34	28.7	16.1
41	28.9	16.9
47	28.9	15.6

LSD(5%) between carbon types = 3 mg/g
LSD(5%) between times = 1 mg/g

[1]Mean of 2 replications w/5 repetitions per rep.

Degradation of 2,4-D Herbicide Sorbed to Activated Carbon. There was little complete 2,4-D mineralization observed in this study. The greatest recovery of initially applied radioactivity was from the liquid (3.5 ± .07%) followed by soil (2.4% ± 0.8%), carbon (0.2 ± 0.1%), and carbon plus 100 ppm alachlor (0.05 ± 0.05%) (Table III). Moreover, the total amount of carbon evolved as CO_2 from the 2,4-D treated samples was less than the non-treated controls. These results imply that microbial activity was inhibited by the presence of 2,4-D, and even more so by 2,4-D plus alachlor.

Table III. Degradation of 2,4-D Sorbed To Carbon After 108 d Incubation at 19 ± 2° C

	DPM's Evolved		mg C-CO_2 Evolved	
Matrix	Net[1]	%[2]	Total	Net[1]
Liquid	27,997(±5,766)	3.5(±0.7)	2869	NA[3]
Soil	18,785(±6,681)	2.4(±0.8)	2821	-281
Carbon	1,511(±531)	0.2(±0.1)	2663	-218
Carbon + Alachlor	3,860(±812)	0.05(±0.05)	2232	-649

[1]Total minus control
[2]Percentage of initially applied DPM's
[3]Not Available

DISCUSSION

We observed varied effects of sorbed pesticides on CO_2 evolution from pesticide-treated carbon. Carbon dioxide evolution from actual spent carbon appeared to be unaffected by the sorbed pesticides while carbon freshly treated with 2,4-D exhibited reduced CO_2 evolution. Pesticide mixtures can either enhance or inhibit microbial respiration (*8*). However, a probable explanation for the varied effects of sorbed pesticides observed in this study concerns the age of the sorbents. It is possible that aging of the actual spent charcoal reduced the biovailability of the sorbed pesticides.

The effect of aging on the biodegradation of soil contaminated with alachlor, atrazine, and metolachlor was described by Felsot and Dzantor (1991) (Proc. EPA\TVA International Workshop on Research in Pesticide Treatment, Disposal, and Waste Minimization, in press). Generally, the degradation of these compounds in aged soils was significantly less than that in soils freshly treated with equivalent herbicide concentrations. The authors attributed the reduced degradation to the limited bioavailability of the herbicides adsorbed to aged soil collected from a pesticide dealership site. The aging effect is apparently due to the pesticide gradually becoming more strongly adsorbed to the interstitial soil pores, which can significantly reduce the amount of pesticide available for biodegradation (*9*).

Limited desorption of pesticides sorbed to actual spent charcoal was also observed by Dennis (*10*). This researcher observed little leaching of seven pesticides when sorbed to activated carbon holding 5.4% of its weight in pesticides. Thus, it is plausible that microbial respiration was not affected by the presence of high concentrations of pesticides on our actual spent carbon because the pesticides were essentially irreversibly bound to the carbon.

This might explain why the amount of CO_2 evolved from the spent charcoal was not affected by the presence of a mixture of pesticides present at high concentrations. However, in the 2,4-D study, the freshly added 2,4-D was potentially more bioavailable. Moreover, high concentrations of 2,4-D can inhibit biodegradation (*11*). Parker and Doxtader (*12*) found that 2,4-D soil concentrations greater than 40 ppm inhibited the microbial degradation of the herbicide. Thus, the increased bioavailability of the freshly added 2,4-D, combined with its high (380 ppm) concentration, could have inhibited microbial degradation in our study. This would explain why our recovery of $^{14}CO_2$ was much lower than that of McCall et al. (*9*) who recovered from 49 to 83% of ring-labelled 2,4-D as $^{14}CO_2$ from various soils after 150 days of incubation.

It is important to note that the amount of 2,4-D that we added to carbon in our study represents a small fraction of the amount of 2,4-D which might be present on actual spent carbon. Dennis (*10*) found that under field conditions, and in the presence of six other pesticides, a 2,4-D ester formulation adsorbs to carbon at a rate of about 7 mg 2,4-D per g carbon. In other words, the 380 ppm 2,4-D that we added to carbon represents only about 5% of the loading capacity of typical carbon.

CONCLUSIONS

In our preliminary studies, we have observed little evidence of significant biodegradation of carbon-bound pesticides. Composting spent carbon with soil after a preliminary chemical treatment (such as with H_2O_2) might enhance degradation and should be evaluated further. However, optimizing conditions for strict biological degradation will be challenging, if not impossible, due to the limited bioavailability, high concentration, and mixture-effects of pesticides sorbed to spent carbon generated in rinsate disposal.

LITERATURE CITED

1. Kobylinski, E.A.; Dennis, W.H., Jr.; Rosencrance, A.B. In *Treatment and Disposal of Pesticide Wastes*; Krueger, R.F., Seiber, J.N.; ACS Symposium Series No. 259; ACS: Washington, D.C., 1984; pp. 125-151.
2. Nye, J.C; In *Pesticide Waste Disposal Technology*; Bridges, J.S., Dempsey, C.R.; Pollution Technology Review 148; Noyes Data Corp.: Park Ridge, New Jersey, 1988; pp. 44-49.
3. Zanitsch, R.H.; Stenzel, N.H. In *Carbon Adsorption Handbook*; Cherermisinoff, P.N., Ellerbusch, F.; Ann Arbor Science Publishers: Ann Arbor, Michigan, 1978; pp. 215-239.
4. Ferguson, T.L.; Wilinson, R.R. In *Treatment and Disposal of Pesticide Wastes*; Krueger, R.F., Seiber, J.N.; ACS Symposium Series No. 259; ACS: Washington, D.C., 1984; pp. 181-191.
5. Anonymous; Summary of National Conferences and Workshops on Pesticide Waste Disposal; OPP/USEPA, July, 1988; 73 p.
6. Wolf, D.C.; Legg, J.O. In *Isotopes and Radiation in Agricultural Sciences*; L'Annunziata, M.F., Legg, J.O.; Academic Press: New York, New York, pp. 100-139.
7. Benedek, A.; In *Activated Carbon Adsorption of Organics from the Aqueous Phase*; McGuire, M.J., Suffet, I.H.; Ann Arbor Science Publishers: Ann Arbor, Michigan, 1980, Vol. 2; pp. 269-297.
8. Stojanovic, B.J.; Kennedy, M.V.; Shuman, F.L., Jr.; *J. Environ. Qual.* 1972, *1*, 54-62.
9. McCall, P.J.; Vrona, S.A.; Kelly, S.S.; *J. Agric. Food Chem.* 1981, *29*, 100-107.
10. Dennis, W.H., Jr.; In Pesticide Waste Disposal Technology; Bridges, J.S., Dempsey, C.R.; Pollution Technology Review 148; Noyes Data Corporation.: Park Ridge, New Jersey, 1988, pp. 44-49.
11. Majka, J.T.; Cheng, H.H.; Muzik, T.J. *J. Environ. Qual.* 1982, *11*,4,645-649.
12. Parker, L.W.; Doxtader, K.G. *J. Environ. Qual.* 1982, *11*, 4, 679-684.

RECEIVED February 20, 1992

Chapter 19

Biological Methods for the Disposal of Coumaphos Waste

D. R. Shelton, Jeffrey S. Karns, and Cathleen J. Hapeman-Somich

Pesticide Degradation Laboratory, Beltsville Agricultural Research Center, Agricultural Research Service, U.S. Depeartment of Agriculture, Beltsville, MD 20705

Two methods have been developed for the disposal of large volumes of spent coumaphos, a livestock tick acaricide, generated annually from cattle dipping operations. The first involves hydrolysis of coumaphos using an organophosphate hydrolase enzyme, followed by ozonation of the hydrolysis products. The resulting ozonation products are readily mineralized by soil microorganisms. The second method involves coumaphos mineralization by a microbial consortium indigenous to one of the vats. The metabolic pathway was partially elucidated and the most important consortium members isolated and characterized. Optimal fermentation parameters were defined. Both methods have been successfully demonstrated at the bench-scale.

Coumaphos [*O,O*-diethyl *O*-(3-chloro-4-methyl-2-oxo-2*H*-1-benzopyran-7-yl phosphorothioate] is used as an acaricide for control of the southern cattle tick (*Boophilus microplus*) and the cattle tick (*Boophilus annulatus*) by the Animal and Plant Health Inspection Service (APHIS), U.S. Department of Agriculture (USDA), in its Tick Eradication Program. Several hundred thousand head of cattle are dipped annually using approximately 60 vats located along the U.S.-Mexican border. The vats contain approximately 15,000 L of coumaphos solution, 0.15-0.30 % active ingredient, in the form of Co-Ral Wettable Powder or Co-Ral Flowable Liquid formulations *(1)*. Vats are emptied, cleaned, and recharged annually, or more frequently depending on the number of cattle treated, sediment accumulation, or loss of efficacy. Cumulatively, approximately 10^6 L of coumaphos waste is generated annually, and is disposed of by pumping into evaporation pits present at each vat-site.

Several strategies were considered for more environmently sound disposal of coumaphos waste. Early experiments had indicated that a *Flavobacterium sp. (2)*, isolated for its ability to hydrolyze organophosphorous insecticides *(3)*, could also hydrolyze coumaphos to chlorferon (3-chloro-4-methyl-7-hydroxycoumarin). This product was readily mineralized via a combination of UV-ozonation and microbiological treatment. In a field trial, 600 L of cattle dip solution was successfully treated by growing the *Flavobacterium sp.* in the solution amended with xylose as growth substrate, ammonium sulfate, and phosphate *(4)*. In a subsequent field trial, however, the *Flavobacterium sp.* failed to grow, apparently due to a deficiency in metal ions *(5)*, which suggested that the growth of *Flavobacterium sp.* directly in cattle dip solutions was not practical. Research was initiated to develop a more reliable method for coumaphos disposal using a combination of enzymatic hydrolysis and ozonation techniques (the enzymatic/ozonation approach).

A second strategy arose from studies to determine the reason(s) for a loss of efficacy in the field, i.e., accelerated rates of coumaphos degradation in several high use vats. These experiments revealed that coumaphos was reductively dechlorinated to potasan [*O,O*-diethyl *O*-(4-methyl-2-oxo-2*H*-1-benzopyran-7-yl) phosphorothioate] under anaerobic conditions and that both potasan and coumaphos were subsequently mineralized under aerobic conditions *(6)*. Research was undertaken to isolate the responsible microorganisms and to elucidate the pathways of coumaphos metabolism, and to determine if these microorganisms indigenous to the vats could be exploited for the disposal of the waste coumaphos solutions (the microbial fermentation approach).

Enzymatic/Ozonation Approach

Studies were conducted with cell-free enzyme extracts to optimize the conditions for coumaphos hydrolysis. Enzyme extracts were derived from cultures of *Flavobacterium sp.*, or recombinant strains of *Esherichia coli (7)* or *Streptomyces lividans (8)* containing the cloned organophosphate hydrolase gene. Due to the very low solubility of coumaphos (50 ppb), experiments were undertaken to assess the effect of a non-ionic detergent (Tween 80) on the rate of hydrolysis. Increasing concentrations of detergent had a dramatic effect on the rate of coumaphos hydrolysis (Table I). Since previous research had indicated that cobalt or copper were stimulatory to rates of organophosphate hydrolase activity with parathion as the substrate *(5)*, experiments were conducted to assess the effect of cobalt sulfate on rates of coumaphos hydrolysis. Increasing concentrations of cobalt also increased the rate of coumaphos hydrolysis (Table I).

TABLE I. Rate of Coumaphos Hydrolysis by Organophosphate Hydrolase

Percent	*First Order Rate Constant*	
Concentration	*Tween*	*Cobalt*
0	0.0048	0.0019
0.001	0.0065	0.0019
0.010	0.0140	0.0023
0.100	0.0220	0.0042

A bench-scale experiment with 76 L of dip vat solution was undertaken to demonstrate the feasibility of the enzymatic/ozonation approach. The pH of the waste solution was adjusted to between 9.5 and 10.0, and 0.05% (v/v) of Triton X-100 was added. Twenty milliliters of a crude enzyme extract (from *Flavobacterium sp.*) containing 19,760 International Units of parathion hydrolase activity was added at time zero. Hydrolysis of coumaphos to chlorferon and potasan to 4-methylumbelliferone was complete within 48 hours (Figure 1). Subsequent ozonation of chlorferon and 4-methylumbelliferone resulted in their complete transformation to numerous unidentified products within 10 hours (Figure 2). Previous research had demonstrated that ozonation products are readily mineralized by soil microorganisms *(2)*.

Microbial Fermentation Approach

Microbial enrichment cultures were started from samples of cattle dip solution using the flowable liquid formulation (42% coumaphos, 58% inert ingredients) as a sole energy and carbon source. After several transfers (10% inoculum), attempts were made to isolate the strains responsible for coumaphos metabolism. Three bacterial strains capable of coumaphos hydrolysis were obtained, but only one strain (B-1) was capable of further metabolism of chlorferon. This strain, however, was not able to utilize

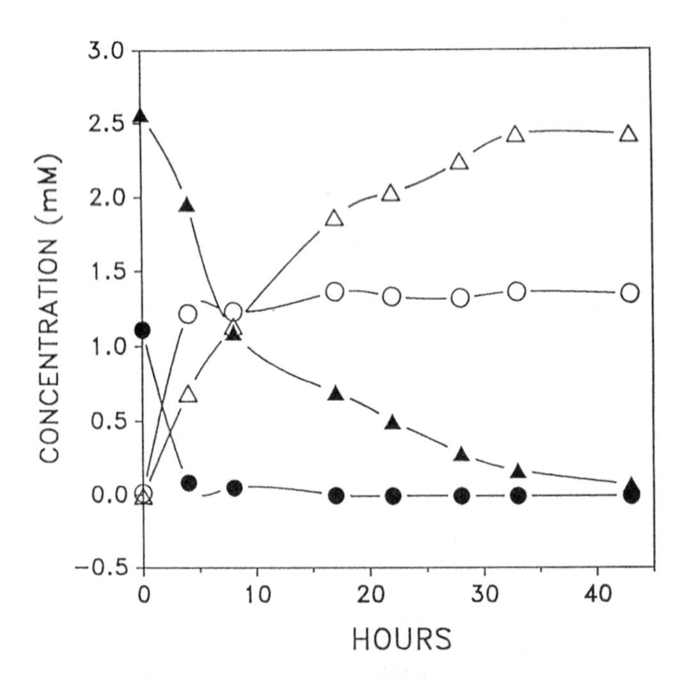

Figure 1. Hydrolysis of coumaphos and potasan in 76 L of cattle-dip waste with organophosphate hydrolase (260 I.U./L) from *Flavobacterium sp.* (▲) coumaphos; (△) chlorferon; (●) potasan; (○) methylumbelliferone.

chlorferon as a growth substrate because one or more chlorferon metabolites were inhibitory to growth. B-1 was able to utilize 4-methylumbelliferone as a growth substrate with no apparent inhibitory effects. Another strain (B-4) was subsequently isolated which, when combined with B-1, resulted in the utilization of coumaphos and chlorferon as a growth substrate (Figure 3). Presumably, B-4 was able to utilize the inhibitory chlorferon metabolite(s) produced by B-1 as growth substrates, thereby removing them from solution and allowing the growth of both strains. Approximately 90% of the chlorferon was mineralized, with about 10% transformed to a metabolite, α-chloro-β-methyl-2,5,4-trihydroxy-*trans*-cinnamic acid, (CMTC) which was degraded by one or more unidentified bacteria (Figure 4) *(9)*. An analagous metabolite (MC) was produced from 4-methylumbelliferone by B-1. The other hydrolysis product of coumaphos and potasan, diethylthioposphoric acid (DETP), was mineralized by a different consortium of microorganisms *(10)*.

Bench-scale experiments with 1 L samples of cattle dip solutions from several vats containing either the flowable liquid or wettable powder formulation were conducted to determine if the microbial consortium could be used to dispose of cattle dip wastes. Solutions were buffered at pH 6.8-7.2 and amended with 100 ppm yeast extract as a nutritional supplement; samples were incubated for 10 to 14 days at 23°C. With the exception of San Andreas and Pinto (mechanical failures) degradation of coumaphos generally exceeded 99% and final coumaphos concentrations were less than 10 ppm (Table II) *(11)*.

Comparison of Approaches

Since neither disposal technology has been attempted at the pilot-scale, precise comparisons are not possible, however, it is possible to make qualitative judgments. The enzymatic/ozonation approach is likely to be more expensive due to both higher fixed and capital costs. Because the enzyme is degraded over time, this approach requires a constant supply of enzyme. Enzyme production costs will be dependent upon the scale of production and the strain used. Organophosphate hydrolase is membrane bound in *Flavobacterium sp.*, but soluble in the recombinant strains *E. coli* and *Streptomyces lividans*, suggesting that production costs may be lower with the recombinant strains due to the ease of enzyme recovery. Alternatively, the development of immobilized enzyme technologies may allow the reuse of the enzyme, thereby lowering costs. However, many technical difficulties remain to be overcome before this would be practical. Other fixed costs would include the operation and maintenance of the ozone generator and ozonation vessel. Capital costs would include the purchase of the ozone generator and ozonation vessel. The trade off for higher cost is greater reliability. The amount of enzyme can be readily changed depending on the coumaphos concentration or other factors effecting enzyme activity. Likewise, the length of time of ozonation can be varied so as to achieve complete destruction of chlorferon.

In contrast to the enzyme/ozonation approach, the microbial fermentation approach is likely to be less expensive but also less reliable. Fixed costs are minimal since fermentation of the contents of one vat results in the production of inoculum for other vats. A source of organic and/or inorganic

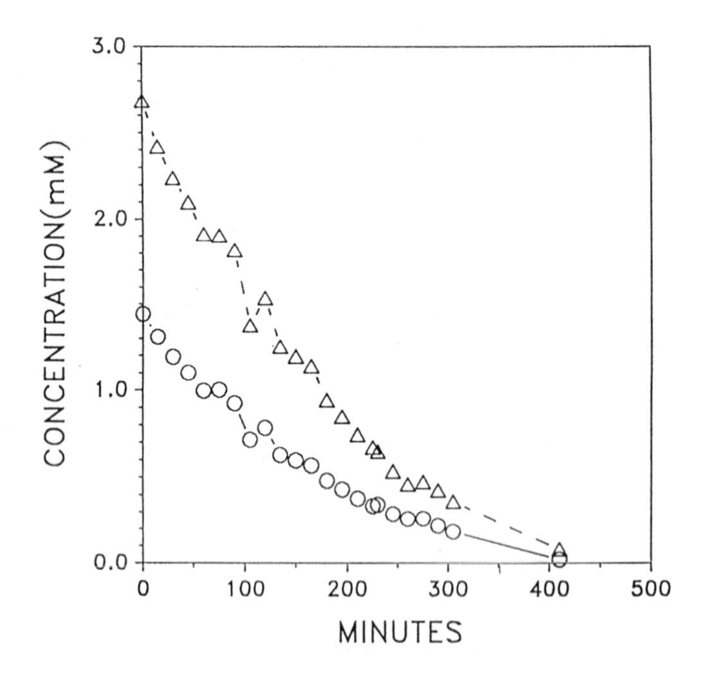

Figure 2. Destruction of chlorferon (△) and methylumbelliferone (○) by ozone, previously treated with organophosphate hydrolase.

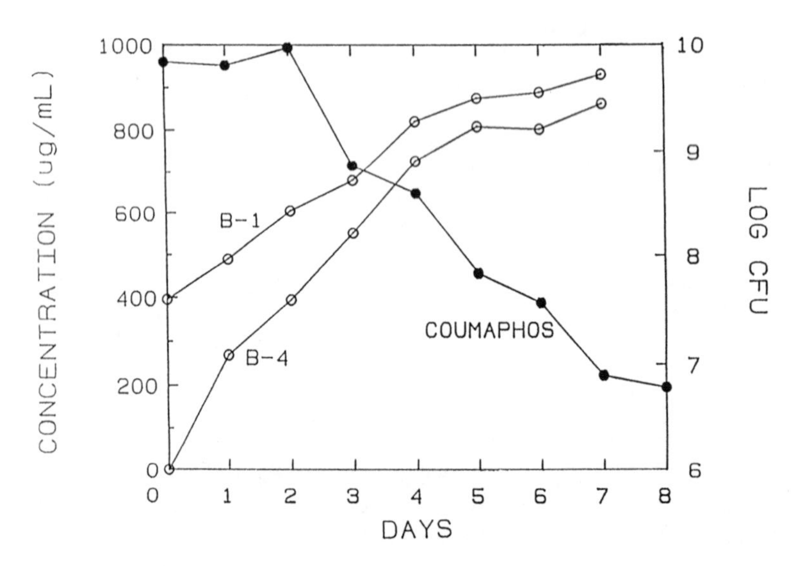

Figure 3. Utilization of coumaphos as a growth substrate by a two-membered consortium consisting of B-1 and B-4. (●) coumaphos; (○) CFU, colony forming units. (Reproduced from *11* Butterworth-Heinemann.)

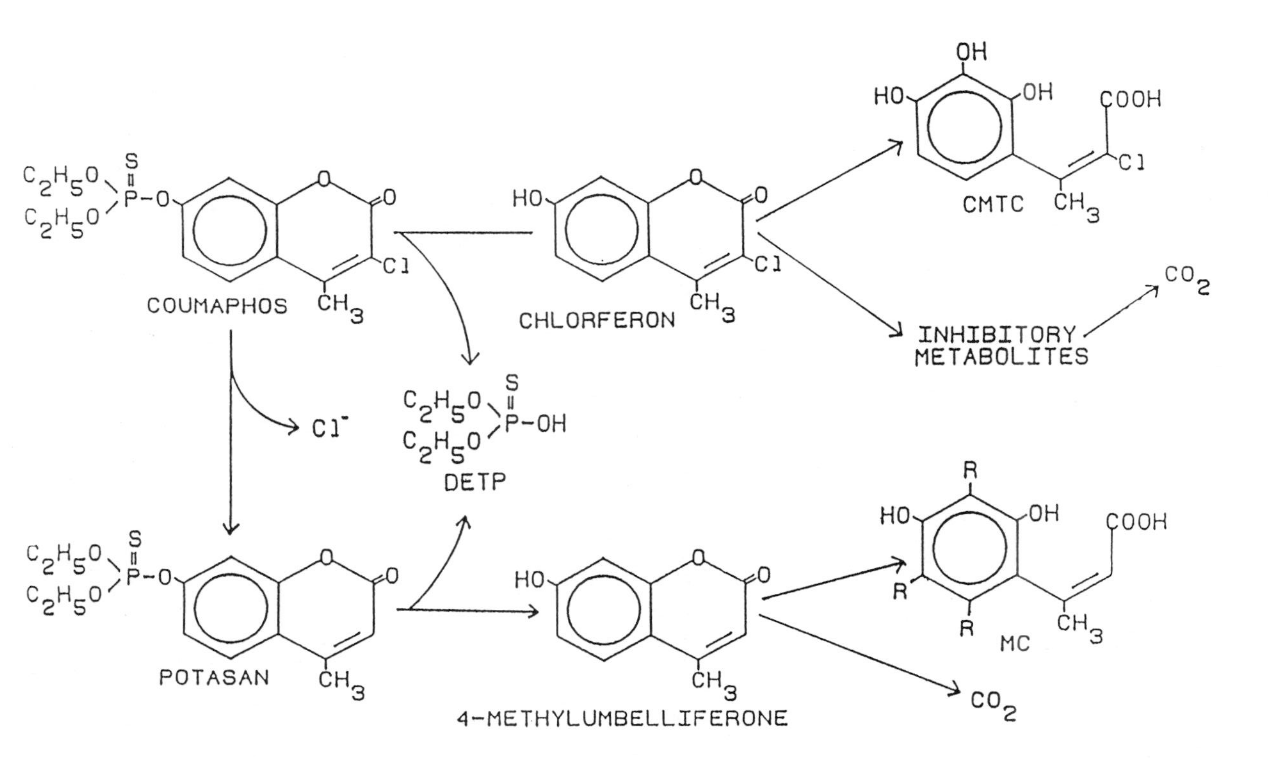

Figure 4. Partial pathway of coumaphos (B-1 and B-4) and potasan (B-1) metabolism. Potasan was produced as a result of reductive dechlorination of coumaphos. DETP was mineralized by a separate bacterial consortium. CMTC and MC were metabolized by one or more unidentified microorganisms. (Reproduced from *11* Butterworth-Heinemann.)

Table II. Coumaphos Degradation in Vat Dip by Microbial Consortium

Vat (Formulation)[1]	*Number of Cattle Dipped*	*Final Concentration (mg/L)*	*Percent Degradation*
Laredo Import (FL)	11,923	3	99.8
Juarez (FL)	32,252	3	99.7
Laredo City (WP)	951	4	99.6
Zapata (WP)	375	4	99.7
Calaboz (WP)	-	4	99.8
San Andres[2] (WP)	387	12	98.8
Pinto[3] (WP)	451	10	97.4

[1] FL = Flowable Liquid, initial coumaphos concentration ca. 0.15%
WP = Wettable powder, initial coumaphos concentration ca. 0.1%
[2] pH probe failed resulting in low pH.
[3] Aeration and agitation failed.

nutrients may be needed, as well as acid and base to control pH, but these amendments and reagents are comparatively inexpensive. Capital costs include a fermentation vessel with aeration system and pH control. Temperature controls are probably not needed since disposal could presumably be scheduled for spring and fall seasons when daily temperatures would be conducive to rapid rates of degradation. Despite the apparent hardiness of the culture, microbial fermentation may be less reliable than the enzyme ozonation approach. Although, in the lab, the method appears to be reliable based on the limited number of vats tested, unknown inhibitors or fluctuations in metal ion composition in some vats may render the method ineffective. At even a low frequency of occurrence, this could result in a loss of cost effectiveness since a back-up method (enzymatic/ozonation) would be required. Further pilot-scale research with both methods will be required in order to answer these questions.

Literature Cited

1. Rogers, B. Animal and Plant Health Inspection Service, USDA, personal communication, 1990.
2. Kearney, P.C.; Karns, J.S.; Muldoon, M.T.; Ruth, J.M. *J. Agric. Food Chem.* **1986**, *34*, 702-706.
3. Sethunathan, N.; Yoshida, T. *Can. J. Microbiol.* **1973**, *19*, 873-875.
4. Karns, J.S.; Muldoon, M.T.; Mulbry, W.W.; Derbyshire, K.; Kearney, P.C. In *Biotechnology in Agricultural Chemistry*; LeBaron, H.M.; Mumma, R.O.; Honeycutt, R.C.; Duesing, J.H., Eds.; ACS Symposium Series No. 334; American Chemical Society: Washington, DC, 1987; pp 156-170.
5. Karns, J.S. Pesticide Degradation Laboratory, USDA/ARS, unpublished data.
6. Shelton, D.R.; Karns, J.S. *J. Agric. Food Chem.* **1988**, *36*, 831-834.

7. Mulbry, W.W.; Karns, J.S. *J. Bacteriol.* **1989**, *171*, 6740-6746.
8. Steiert, J.G.; Pogell, B.M.; Speedie, M.K.; Laredo, J. *Bio/Technology* **1989** *7*, 65-68.
9. Shelton, D.R.; Hapeman-Somich, C.J. *Appl. Environ. Microbiol.* **1988**, *54*, 2566-2571.
10. Shelton, D.R. *Appl. Environ. Microbiol.* **1988**, *54*, 2572-2573.
11. Shelton, D.R.; Hapeman-Somich, C.J. In *On-Site and In Situ Bioreclamation*; Hinchee, R.E.; Olfenbuttel, R.F., Eds., Butterworth-Heinemann: Boston, MA, 1991; pp 313-323.

RECEIVED May 12, 1992

Chapter 20

Site Assessment and Remediation for Retail Agrochemical Dealers

Christopher A. Myrick

National AgriChemical Retailers Association, 1155 Fifteenth Street N.W., Suite 300, Washington, DC 20005

Many retail agrichemical dealers have slowly, and in most cases inadvertently, contaminated the soil and water at their dealerships with varying levels of pesticides since the use of these products began to increase in the early 1950s. Because of this contamination and various other forces, one of the most uncertain and ominous issues now facing the agrichemical industry is the cleanup of contaminated soils and water at retail dealerships. Currently dealers have very limited and expensive options available to them for cleanup of contaminated soil and water. For example, if the soil contains pesticide waste regulated under the Resource Conservation and Recovery Act (RCRA), estimates show that it could cost from one to five million dollars to cleanup a single facility. In addition, a large data gap exists with respect to the remediation of chemically contaminated soils and water at retail facilities, prohibiting regulators from reviewing the viability of new and more cost effective cleanup procedures. Development of new and existing technologies that will allow for low cost and user friendly site assessment and remediation at retail dealerships will save the United States agricultural industry billions of dollars over the next two decades.

Issues Forcing Dealers to Explore Research Options

Dealer Insurance. Insurance coverage is the most prevalent factor forcing the retail segment of the agrichemical industry to carefully assess the site contamination issue. Currently, it is very difficult or impossible for dealers to acquire insurance coverage for the cleanup of sudden spills, much less a series of accidental spills that have occurred over a prolonged period of time. According to a recent NARA survey, over 60 percent of retail dealers have no short or long term environmental pollution coverage.

0097–6156/92/0510–0224$06.00/0

The insurance issue has become even more critical recently due to court cases involving claims for the cleanup of contamination resulting from long term chemical releases. When dealers have filed insurance claims, insurance companies have gone to court arguing that their "sudden and accidental" spill coverage does not insure dealers for the cost of remediating the contamination resulting from long term or existing accidental chemical releases.

Because of pending and precedent setting decisions about insurance coverage of long term chemical contamination, the agrichemical industry must immediately take a lead role in developing economically viable site assessment and remediation technologies. Through the development of these technologies, dealers can begin conducting voluntary audits of their facilities and begin remediating when needed. In turn, site assessment and completed remediation activities will allow dealers to prove to insurance companies that their facilities are environmentally sound, allowing them to obtain adequate insurance coverage at reasonable premiums.

Lender Liability. Lending institutions are becoming increasingly concerned about their legal and financial obligations should a dealer that they are financing become subject to regulatory enforcement requiring soil and water remediation. In fact, several congressional hearings have taken place on this subject during the past few years because of its tremendous ramifications.

Considering that many agrichemical dealers are not financially capable of covering the huge cost associated with cleanup, the potential for them to forfeit contaminated property with the accompanying liability is a real possibility. The long term financial integrity of agrichemical dealers is in jeopardy unless economically viable site assessment and remediation procedures are discovered, allowing lending institutions to re-enter the retail dealer market.

1988 FIFRA Amendments. Another emerging concern of dealers is the implications of the Federal Insecticide, Fungicide, and Rodenticide Act Amendments of 1988 (FIFRA '88). Under the mandate of FIFRA '88, the U.S. Environmental Protection Agency (EPA) is in the process of drafting extensive regulations that will require dealers to invest substantial amounts of money to build dikes, rinse pads, and perform other costly renovations to their facilities that will prevent further contamination.

Many dealers question the value of making large investments to comply with the new FIFRA '88 regulations on sites that may have to be excavated due to previous contamination. In light of this, the agrichemical industry along with the EPA must conduct research and development activities that will clarify misunderstandings and insure dealers that their FIFRA '88 investments will not have to be torn down for excavation. These concerns, as well as the impending reauthorization of RCRA, force dealers to make decisions about assessing and remediating their facilities.

Current Regulatory Climate

Several states have initiated the task of assessing and remediating agrichemical dealerships. From preliminary reports, however, more questions about remediation and site assessment have been raised than have been answered. The following discussion will review some of the more pressing issues that regulators face.

Overlap of FIFRA, CERCLA, RCRA. A tremendous amount of confusion exists regarding the treatment of suspended and/or canceled pesticides under FIFRA, RCRA, and the Comprehensive Environmental Response, Compensation, and Liability Act (CERCLA), especially if different pesticides regulated under these different laws are mixed in soil matrices. For example, soils that are contaminated with prohibited waste under RCRA are subject to more stringent land disposal restrictions. These restrictions require certain pesticide waste to meet expensive technology-based treatment standards, e.g., incineration before they can be land disposed. In light of this example and others, EPA must consider new disposal rules that will give dealers more options when treating soils contaminated with pesticides regulated under various federal and state laws.

These options may include allowing for the land application of pesticide-contaminated waste at FIFRA-approved label rates. Under this approach, contaminated soil would be viewed as a "new inert diluent," and as such, it would be considered a carrier for the active ingredient(s) within the pesticide formulation. Using the land application option, the appropriate FIFRA-approved label rate would be based on a target pesticide, i.e., the one of highest concentration. Once this rate is determined, the pesticide-contaminated soil may then be applied or redistributed at or below its "label rate" for the type of crop grown on the receiving field. Land application will allow natural photolysis and biodegradation to break down the hazardous constituents (Vorback, J., U.S. EPA/OSW, personal communication).

Since the land application approach is legal under RCRA because it represents recycling and reuse of a product, the EPA also believes that it should be legal under FIFRA because the target pesticide(s) are being applied according to label rates on label-approved crops. However, success of this method rests heavily on the specific location and pesticide mixtures in the soil.

Should the option allowing land application of pesticide-contaminated soil not be acceptable under RCRA, contaminated soils would then have to be remediated following established RCRA or special state guidelines which are very expensive. Because of this, a special disposal and treatment option currently being considered by the EPA would come into play.

If promulgated treatment standards set forth under RCRA cannot be met (*1*), the second option involves granting a special soil and debris variance allowing for the use of alternative treatment standards, which are similar to those currently available under CERCLA. At present, a treatment standard is based on the performance of the best demonstrated available technology (BDAT) to treat the waste. Compliance with performance standards may be ascertained by determining the concentration level(s) of the principal organic hazardous constituents (POHCs) in the waste, treatment residual, or the extract of the waste or treatment residual.

Alternative treatment standards are now granted under CERCLA in the form of a site specific variance (*2*) or a full-scale variance (*3*). The primary obstacle to granting a site-specific variance from a RCRA prescribed treatment standard is the requirement of the petitioner to demonstrate that the physical (or chemical) properties of the waste differ significantly from the waste analyzed and used to develop the treatment standard; therefore, the waste cannot be treated to RCRA specified levels (*4*).

To meet the above requirement, the agrichemical industry must produce research data to show that treatment levels cannot be met using prescribed treatment technology because the physical (or chemical) properties of mixed waste in the soil of agrichemical dealerships varies significantly from the waste analyzed to develop the prescribed treatment levels. In many cases, the prescribed treatment technology is incineration; however, the technologies used under a variance must be economically favorable for dealers to use. An example of an affordable technology would be treating soil with a microorganism that breaks the pesticide waste down to a prescribed toxicity level so the waste can then be land applied.

Site Assessment. Standard criteria for site assessment have not been developed and adopted nationwide. Primary problems encountered in developing a single site assessment method are the separation and identification of inert waste, hazardous waste, special state-listed waste, and other waste constituents in soil and debris matrices. Beyond the problems that mixtures of pesticide, fertilizer, and other waste present, certain legal questions arise when the specific waste generation source cannot be clearly defined during site assessment.

For example, the EPA has determined that residual pesticides washed off application equipment are not wastes under current federal laws. Consequently, soil contaminated with pesticides as a result of the discharge of exterior rinse water is not a RCRA waste (even though one or more of the pesticides may have accumulated to levels that exceed the numerical values for RCRA waste which are specified in 40 CFR). In addition, waste from commercial products that are mixtures of a listed pesticide and other active ingredients are not regulated as RCRA waste (*5*) unless they meet certain criteria (*6*). Even though these wastes are not technically regulated under RCRA, a legal question arises when the waste migrates into the ground water; does the soil then have to be treated and disposed of using RCRA remediation procedures, or are other remediation methods acceptable?

This example is just one of many different scenarios that may arise when dealing with hazardous waste mixtures in soil. Overcoming the site assessment problem would be helped by establishing a comprehensive site assessment data base that would give dealers a quick summary of the hazardous constituents present and tell them what remediation methods are legal and best suited for the cleanup of their specific location.

Establishing Remediation Trigger and Cleanup Levels. One concern that has been raised by regulators involved in the remediation process is the need to establish uniform remediation trigger levels that provide flexibility. (The trigger level is the contaminant concentration in the soil that would require remediation). Flexibility is needed because of site variations including pesticide types, soil and bedrock properties, ground water depth, surrounding land use, and proximity to wellheads.

One example that can be used to analyze the process of establishing a trigger level comes from the State of Wisconsin (7). Wisconsin regulators have established three approaches for setting remedial action trigger levels. An early method involved setting a single trigger level for all pesticides. In areas where the concentration of one or more of the pesticides in a soil sample exceeds this level, the responsible party removes all soil, gravel, and debris from the area. These types of trigger levels have ranged from 1 to 10 ppm and were based on pesticide concentrations expected to be found in the surface soils of agricultural fields following normal rates of pesticide application. This method of defining trigger levels was used in most Wisconsin cases. Wisconsin officials felt that pesticide residue concentrations below this trigger level would not result in adverse environmental effects. Residue concentrations in soil samples were compared with the trigger level regardless of the sample depth (although most samples were collected from the soil surface).

Recently, Wisconsin adopted a new approach in which two trigger levels are set for a facility. One level is set for the surface soil samples and a second, lower level is established for all subsurface soil samples. Lower trigger levels for subsurface samples were needed because little biotic activity or organic matter exists at such depths. Consequently, limited biodegradation occurs below the soil surface layer, and pesticides found at depth are more likely to reach ground water. Setting dual trigger levels seems to be a reasonable approach to remedial action and assessment.

A third method for setting remedial action trigger levels has also been utilized in Wisconsin. This method incorporates the dual trigger level method discussed above plus the establishment of a trigger level based on the cumulative concentrations of all pesticide compounds found at all depths at any one location. This combined trigger level approach addresses both the potential of pesticide interactions and the possible effects of pesticides present at various depths.

Although the Wisconsin activities address the issue of trigger levels, several states are still trying to resolve the question of where to establish trigger levels. In light of so many unanswered questions and the direct effect that the establishment of trigger levels can have on the cost of cleanup, a tremendous need exists for research and development in this area. Closely linked to the creation of trigger levels is the establishment of post cleanup levels, the pesticide concentrations to which soil and water must be cleaned to be considered safe.

Research Needs

After reviewing the current issues that are moving remediation concerns forward and assessing the current laws that may regulate the assessment and remediation of a facility, clearly a tremendous amount of research needs to be conducted. This research should be aimed at providing environmentally sound and economically efficient site assessment and remediation tools to retail dealers.

Following is a review of suggested research and development needs that have been highlighted by NARA to remediate currently contaminated retail facilities.

Research and Development in Site Assessment.

Development of preliminary site assessment procedures that can be undertaken by a retail dealer. Today, an environmental consultant hired by a dealer to do a site assessment must report any contamination found to State authorities. This requirement is currently prohibiting many dealers from moving forward with any corrective action because of the fear of financial ruin due to the high cost of remediation. The development of visual as well as preliminary sampling procedures that can be carried out by retail dealers will help the industry move forward in discovering and correcting contaminated sites by reducing the cost of site assessments.

Specific research needs include:

1. Development of guidelines that can be used in historical records check of a facility. These guidelines would help pinpoint specific practices or evidence that would indicate high probability of contamination.
2. Development of guidelines for visual site assessment procedures based on vegetative state, soil decolorization, proximity to wells and streams.
3. Establishment of priority constituents commonly detected in soil and water at retail agrichemical facilities.

Development of sampling technology that is centered on the unique characteristics of agrichemical facilities. Sampling soils and water that may be contaminated with pesticides, fertilizers, and solvents is a very complex task. At this time, there is a minimal amount of research going on in this area. Much more research must be carried out to gain a better understanding of how to sample soils with high concentrations of mixed pesticide wastes. Sampling technologies will also have to consider soil type, hydrology, and potential for constituent(s) to leach.

Specific research needs include:

1. Establishing the validity of a variety of different extraction procedures and their relationship to constituent leaching potential.
2. Development of low cost sampling procedures that are workable on soil containing a complex mixture of pesticides.
3. Development of standardized analytical procedures that would simulate and predict leaching of agrichemicals by actually testing the soils from contaminated sites. The procedures may be generically comparable to the Toxicity Characteristic Leaching Procedure (TCLP).(*8*)

4. Development and validation of enzyme linked immunosorbent assay (ELISA) field test kits for evaluating pesticide contamination in soil and water.

Development of remediation trigger levels based on risk assessment. It has been discovered through previous site assessment activities that individual agrichemical facilities possess unique characteristics. These unique characteristics in many cases may require that regulators establish trigger levels on a facility-by-facility basis. Individual trigger levels must be set conservatively because of the limited amount of information currently available regarding the risk presented by single or multiple waste mixtures in different soil and hydrogeologic situations. The development of trigger level standards for remediation is of utmost importance to the entire agrichemical industry cleanup initiative.

Specific research needs include:

1. Assessment of existing unsaturated-zone transport models for contaminants pesticides spill sites.
2. Development of risk assessment models that would accurately assess different constituents based on actual sampling, predicted leaching potential, and individual facility characteristics.
3. Development of remediation trigger levels from model predictions. These predictions would be based on health hazards of waste constituent(s) and then potential to leach into ground water at each individual location.
4. Development of computer based expert systems that would aid in making trigger and cleanup level decisions based on relevant data.

Development of micro-economic cost analysis formula that can be used on a site-by-site basis. The costs of site assessment activities are of great importance to the goal of cleaning up contaminated sites. Most agrichemical dealers are small businesses and typically operate on a small margin that leaves little capital for site assessment and remediation costs.

The development of a micro-economic cost analysis formula that may be used by agrichemical dealers, consultants, and regulators would be of great benefit in determining whether a particular facility is financially capable of the recommended cleanup. This formula would also be very helpful when cleanup costs must be spread over a prolonged period of time.

Specific research needs include:

1. Development of site assessment cost data base that takes into consideration multiple site assessment technologies based on individual site characteristics.
2. Development of a linear programming model which considers basic facility characteristics and gives lowest cost site assessment procedure based on those characteristics.

3. Development of financial data base which will be used to give lowest site assessment technology and individual facility financial capability.

Research and Development in Site Remediation.

Development of Soil Cleanup Levels. Establishing soil and water cleanup levels that mitigate risk are important to the cleanup objectives of the agrichemical industry and regulators. Although in many cases water cleanup levels may already be specified through Maximum Contaminant Levels and Health Advisories, cleanup levels for soil are much more ambiguous.

Some basic research is already being conducted regarding the establishment of soil cleanup levels based on contaminant concentration, leaching potential, soil characteristics, and hydrogeology. However, more research is needed to gain a more accurate assessment of the risk that different mixtures and concentrations of contaminants present under various conditions.

Specific research needs include the development of attainable soil cleanup levels (thresholds) that take into consideration individual constituents and complex mixtures. These cleanup levels must meet risk/benefit analysis criteria that take into consideration a facility's ability to finance cleanup as well as the actual hazards that the wastes present.

Development of Low Cost Remediation Technology. According to a review of literature on remediation technology, not much is known about the workability of current remediation technology as it relates to agrichemicals. From preliminary cost estimates of technology now being used, it is evident that new in-situ low cost technologies must be developed in order for the agrichemical industry to cleanup contaminated sites while remaining economically viable.

Because of the wide variety of chemical mixtures present at agrichemical facilities, researchers must take into consideration the regulatory implications of technological innovations. For example, waste regulated under RCRA cannot be treated in the same manner as waste that is not regulated under RCRA or special laws. Research must be directed at finding low cost remediation technology that can be carried out by the dealer or consultant under these various laws.

Specific research needs include:

1. Development of a broad base of remediation technologies that are applicable to various sites and can be carried out by the agrichemical dealer under regulatory agency supervision. These remediation technologies must be proven capable of attaining the cleanup objectives of regulatory agencies while remaining low in cost.
2. Development of a remediation technology data base that takes into consideration laws governing remediation activities.
3. Development of low cost land-spreading technology that can be used in degrading contaminated soil and water.

4. Development of in-situ soil composting technology that can be used in decontaminating high concentrations of pesticide waste.
5. Development of portable degradation tanks for on-site treatment of pesticide contaminated soil and water.
6. Evaluation of the feasibility of decontaminating ground water by applying it to farmland though spray irrigation.

Development of Micro-Economic Cost Analysis That Can Be Used on a Site-by-Site Basis. As is true for site assessment, the development of micro-economic cost analysis formula for the remediation of agrichemical facilities would be very helpful in projecting the short and long term costs of remediation activities. Understanding remediation cost and weighing different remediation methods based on cost would be extremely helpful in insuring the continued financial security of the entire agrichemical industry.

Specific research needs include:

1. Development of a data base that contains remediation cost information base on contaminants, leaching potential, soil and water cleanup objectives, and cleanup technology.
2. Development of computer-based "expert systems" that can be used by individual dealers, regulatory officials, leading institutions, and insurance companies in determining the financial capability of a facility to successfully carry out remediation activities.

Conclusions

Considering the various circumstances surrounding the remediation issue, neither time nor technology is working in favor of the retail agrichemical community. To address the situation in an environmentally sound and economically viable manner, research and development activities must be conducted and completed in the near future. Successful completion and adoption of the new technologies that will emerge from the research activities will insure the continued viability of the agricultural sector while protecting our natural resources.

Literature Cited

*1*U.S. EPA (1989a). Obtaining a Soil and Debris Treatability Variance for Remedial Actions. Superfund LDR Guide #6A. OSWER Directive: 9347.3-06FS, July 1989.

*2*Fed. Regist. 53, Aug. 27, 1988, p. 31221.

*3*Resource Conservation and Recovery Act, U.S. Code Title 40, Section 268.44, July 1, 1989.

*4*Resource Conservation and Recovery Act, *U.S. Code* Title 40, Section 268.44, 1988 ed.

*5*Resource Conservation and Recovery Act, *U.S. Code* Title 40, Section 268.44, 1988 ed.

*6*Resource Conservation and Recovery Act, *U.S. Code* Title 40, Section 261.33, 1988 ed.

*7*Resource Conservation and Recovery Act, *U.S. Code Title* 40, Part 260 Subpart C [Section 261.20], 1988 ed.

*8*Fed. Regist. 55, March 29, 1990, p. 11798.

RECEIVED May 21, 1992

Chapter 21

Agricultural Chemical Site Remediation and Regulations

Greg Buzicky, Paul Liemandt, Sheila Grow, and David Read

Agronomy Services Division, Minnesota Department of Agriculture, 90 West Plato Boulevard, Saint Paul, MN 55107

Agricultural communities, industry and states need to address a serious environmental problem that has developed as a result of modern agricultural practices over the last thirty years. Accidental and incidental agricultural chemical spillage that occurred at a variety of sites is impacting, or has the potential to impact, ground and drinking water in many rural communities. For many businesses and regulatory agencies, difficult financial and policy decisions must be made to address an issue that for many years was unknown or ignored.

Preliminary results from a survey of state agricultural agencies by the State FIFRA Issues Research Evaluation Group's (SFIREG) Working Committee on Ground Water Protection and Pesticide Disposal demonstrated the national scope of contamination problems at agrichemical dealerships *(1)*. Of 26 states responding, 21 indicated an awareness of severe environmental impacts resulting from agrichemical handling sites. A total of 720 sites were reported with only five states (Minnesota, California, Florida, Michigan and Wisconsin) accounting for 82% of the sites. Five hundred and twenty sites were involved with soil or ground water remediation with the same five states accounting for 57% of the total. This survey indicated that regulatory agencies are aware of the problem but few are actively addressing the issue.

Minnesota, which has a program to address the issue, listed the most sites at 200. In Minnesota the sites are divided into two main types: sudden releases and long-term contamination. Approximately 225 sudden releases are reported each year with most cleaned up relatively quickly. The approximately 100 long-term contamination sites have been reported through a variety of mechanisms like self-reporting, routine inspections, or complaints.

0097–6156/92/0510–0234$06.00/0

Background

The introduction and use of pesticides and fertilizers played a significant role in increased crop production in the United States. Since the 1960's, modern intensive agriculture has relied on agricultural chemicals as an integral part of crop management. Concerns about contamination of water resources from the normal, registered use of pesticides in the late 1970's and early 1980's resulted in the Environmental Protection Agency (EPA) and several states developing nonpoint assessment and management programs *(2)*.

Debate regarding sources of ground water contamination from targeted state surveys resulted in EPA's development and implementation of a statistically designed, National Pesticide Survey of rural domestic wells and community water systems. The national survey confirmed the findings of various state surveys; some certain pesticides in certain hydrogeological situations could impact ground water as a result of normal use. However, left unanswered was the extent and impacts of pesticide and fertilizer mixing, loading and handling sites. Prior to the survey, there was significant debate about sources of contamination and much of it revolved around the relative importance of "point sources" from agricultural chemical mixing, loading and handling sites. While spills have been historically addressed by the industry and regulatory agencies, very little knowledge existed as to the environmental impact on sites where accidental and incidental spillage of pesticides had occurred. This issue has now become a focal point of concern as information regarding the impacts has developed.

Pesticides are required by state and federal law to be used, stored, and handled in accordance with language on the pesticide label. The specific label instructions prohibit off-target application of pesticide products. Nevertheless, accidental and incidental spillage has commonly occurred at mixing and loading sites. Often spillage would result from routine activities such as mixing pesticide sprays, washing field equipment, washing and draining of spray tanks and impregnation of fertilizers with pesticides. In addition, illegal spillage or dumping of pesticides and sloppy handling of pesticides also occurred, although the true extent of this type of misuse was not known. Early discussion of these concerns in the mid 1980's focused on the potential impacts on ground water contamination and was largely discussed because of the local and state survey efforts that recognized the potential impact.

To provide baseline information on the occurrence and extent of agricultural pesticides in ground water and drinking water, The Minnesota Departments of Health and Agriculture conducted cooperative surveys of water wells from 1985 to 1987 for selected pesticides *(3)*. The widespread occurrences of pesticides, primarily atrazine, in ground water in sensitive hydrogeologic areas was a result of normal pesticide use, but some of the detections resulted from accidental or incidental spillage, backsiphonge, or other types of point sources.

In a paper discussing the impact on ground water by distributors of agricultural chemicals, Victor *(4)* concluded that the principal potential sources of ground water pollution in the San Joaquin Valley by agricultural chemicals included operations of local distributors. Victor also stated that unless the pesticide spillage was chronic and/or involved large amounts, the impact on the quality of soils and ground water was generally small. The author also acknowledged that the potential extent of the ground water problems associated with these sites was not well understood.

In an overview of the impact of pesticides on ground water, Hallberg *(5)* discussed "quasi point sources" associated with agricultural chemical facilities where pesticides in the range of formulation concentrations were found in pools of water in mixing and handling areas. Concentrations of pesticides as high as 100 ppb were detected in nearby shallow drinking water wells. The author also concluded that with over 1500 agricultural chemical facilities in Iowa the potential impact of spills may be widespread.

Agricultural Chemical Facility Sites

No two pesticide mixing, loading, and handling sites are the same. The facilities vary due to a large number of factors. Some of these factors include the types of products handled at the facility, the size of the facility, the hydrogeologic site, the construction and type of containment, and the potential ground or surface water receptors of contamination. Accordingly, it is difficult to describe a "typical" accidental and incidental agricultural chemical spill site.

Generally, there are two classifications of pesticide releases: sudden releases and chronic releases. Sudden releases most frequently occur in the spring and early summer as agricultural chemicals are being handled and applied. Storage tank, leaks, spray tank tipovers, hose breaks, and transport accidents can release hundreds or thousands of gallons of product. These types of releases can be relatively easy to cleanup provided they are reported and responded to immediately. Since the area of the accident and the product involved are usually readily identifiable, the contaminated soil can be excavated, analyzed, and utilized as originally intended at relatively low cleanup cost. Cleanups are relatively easy if the release occurs within a modern containment structure.

A second type of sudden release which is significantly more serious, expensive, and complicated are fires that occur at agricultural chemical facilities. Because large quantities and many types of pesticides are usually involved, it is difficult and expensive to manage the fire debris. Complicating the problem is the actual management of the fire. Firewater may spread the surface soil contamination, cause ground water contamination, and runoff to surface water either directly or through sewers. Cleanups often are expensive and range from hundreds of thousands to millions of dollars depending on the size of the fire, the facility, and the costs incurred in the disposal of the debris.

Accidental and incidental agricultural chemical spill sites were reviewed by Habecker *(6)* and found to be a major point source for ground water contamination by pesticides. In the twenty sites examined in Wisconsin, 17 different pesticides were found in the soil; and 19 different pesticides were found in the ground water. Soil contamination was examined by evaluating various activity areas; the most severely contaminated places involved acute spill areas, burn piles, and pesticide impregnated fertilizer loadout areas. Alachlor, atrazine, cyanazine and metolachlor were most frequently found in the soil. Multiple pesticides were frequently found in ground water and often exceeded health advisory levels.

Sites evaluated in Minnesota reflected trends similar to those in Wisconsin. Individual activity areas demonstrated soil contamination that generally decreased with depth. Ground water impacts were a function of the hydrogeology of the area as well as the nature of the specific compounds. Frequently, multiple detections of pesticides were found in ground water.

In addition to the lack of investigative and agricultural chemical cleanup technologies, the remediation of accidental and incidental contamination of soil presents a challenge for responsible persons and regulators due to the current myriad of federal and state regulations. The regulations were not designed to address remediation of accidental and incidental agricultural chemical spill sites. Application and interpretation of legislation and regulations are complex and subject to debate. In any case, they are a challenge to practical and efficient remediation of soil and ground water. Three federal laws, all administered by the EPA, pertain to agricultural chemical contamination site remediation and in particular to land application of pesticide-contaminated soil.

Federal Insecticide, Fungicide and Rodenticide Act (FIFRA)

Most recently amended in 1988, FIFRA's purpose is to regulate the registration, use, manufacturing, and disposal of pesticides. Pesticides are recognized as potentially toxic chemicals deliberately applied for beneficial purposes into the environment. Accordingly, FIFRA requires EPA to register each pesticide for specific uses that do not pose unreasonable risks to human health or the environment.

As a component of the regulatory process, each pesticide product, which includes the active ingredient and inerts, must be applied in accordance to its product label. The label is developed and approved by EPA and defines the legal restrictions applicable to an individual product regarding its use, storage, handling, and disposal. Included in the label restrictions is information regarding maximum application rates, target sites, and crops. EPA may cancel or suspend the use of a pesticide if the pesticide presents unreasonable adverse effects on human health or the environment.

While the EPA administers FIFRA at the federal level, most states administer FIFRA in addition to their respective state pesticide control laws. State laws can not be more permissive then FIFRA. Under FIFRA and traditional state pesticide control laws, persons responsible for spills and other environmental contaminations are responsible for the costs associated with cleanups. These costs necessarily include assessment and investigation in addition to remediation.

Amendments to FIFRA direct the EPA to establish procedures for storage, transport and disposal of containers, rinsates, or other material used to contain or collect excess or spilled pesticides. Part 165, Section 19 of the 1988 amendments includes Subpart C which addresses containment and cleanup of spills and leaks.

Because the factors important to the degradation process are often not present in the subsoil at the spill site and migration of the spilled pesticides must be prevented, excavation of the contaminated soil is the only viable remediation method. Land application of the excavated soil is frequently considered and utilized. Land application of the contaminated soil allows for the exposure of the pesticide like oxygen, microbes, sunlight, and temperature that increase the rate of pesticide degradation. In addition, application of the pesticide-contaminated soil to an appropriate site for reuse at or below the application rate provides an economic and environmental advantage to other treatment or disposal options.

With respect to land application of contaminated soil, FIFRA regulations mandate that pesticides are applied according to label instructions. Only pesticide products currently (or within a time frame allowed by law) labelled have legal uses. If soil is interpreted as a diluent or carrier, land application of registered pesticides can be utilized. This option is restricted to currently registered products. Because of the accelerated EPA pesticide registration process currently underway, and the anticipated reduction in pesticide products, land application could become a more limited option in the future. Soils contaminated with a cancelled pesticide currently have limited options for cost effective remediation.

Resource Conservation and Recovery Act (RCRA)

RCRA is the primary law regulating the generation, transport, and disposal of hazardous wastes. Hazardous wastes are identified by procedure, either through listing or characteristic. The Code of Federal Regulation, 40 CFR Part 261, identifies hazardous waste; the code offers a list of compounds that are hazardous wastes, describes certain hazardous characteristics (toxicity, ignitability, reactivity, and corrosiveness), and describes the tests for determining whether a waste is hazardous.

RCRA regulations may be pertinent to accidental and incidental agricultural chemical spill site remediation. Sites with contaminated soils containing a listed hazardous waste, in addition to a pesticide product, may fall under RCRA regulations. Cancelled pesticides, which have no currently allowable use, present another potential regulatory situation under the jurisdiction of RCRA. It is anticipated that the number of cancelled pesticide registrations will increase dramatically as a result of the re-registration process mandated by EPA in the 1988 amendments to FIFRA.

Listed RCRA hazardous wastes generally do not have "de-minimis" values assigned to each constituent of the list and therefore are assumed to pose hazards to human health and the environment at any concentration. "De-minimis" values are for the purpose of this discussion concentrations below regulatory concern. The names of some pesticide products or active ingredients in pesticide formulations are listed in RCRA List P, Acutely Hazardous Commercial Products, and List U, Toxic Commercial Products (e.g., List P contains aldrin, dieldrin, endrin, and parathion; List U contains chlordane technical, formaldehyde, lindane, and mercury).

Soil that is contaminated with pesticides that are listed hazardous wastes are not considered hazardous waste if the pesticide is applied to land as a part of the normal use. Soil that is contaminated by spilled pesticides that are listed hazardous wastes may need to be treated as hazardous waste *(7)*.

With respect to land application of pesticide-contaminated soils containing hazardous wastes, the Hazardous and Solid Waste Amendments (HSWA) to RCRA include specific land disposal restrictions (LDRs). LDRs are restrictions that apply to land disposal of RCRA wastes that are "placed" in (but not limited to) a "landfill, surface impoundment, waste pile, injection well, land treatment facility, salt dome formation, salt bed formation, underground mine or cave, and concrete bunker or vault (RCRA 3004(k)" *(8)*.

Treatment standards (concentration levels) are specified for restricted wastes. Pesticides that are restricted wastes are listed in 40 CFR Part 268, Appendix III *(9)*. Treatment standards are not "de-minimis" values and do not mean that levels below the treatment standards are not hazardous wastes, but rather the treatment standard must be met before disposal. Treatment standards are based not on health considerations but on the best demonstrated available technology. Dilution of a waste is prohibited unless the dilution is part of a treatment process (e.g., addition of acid to neutralize a base).

Land disposal restrictions also provide for compliance options in order to allow the development of alternative treatment technologies *(10)*. Various compliance options include a treatability variance, an equivalent treatment method petition, a no migration petition, and delisting. Delisting applies only to listed hazardous wastes. Hazardous waste as defined by characteristic may be treated so that it no longer exhibits the hazardous property. Providing a specific de-minimis concentration for each hazardous waste would allow practical treatment alternatives.

Comprehensive Environmental Response Compensation and Liability Act (CERCLA)

The primary cleanup mechanism utilized is the Comprehensive Environmental Response Compensation and Liability Act (CERCLA), also known as Superfund, which was passed by Congress in 1980 to address past waste disposal practices. Superfund authorized the EPA to: (1) respond to emergencies involving release of hazardous substances, and (2) cleanup highly contaminated sites that threatened public health or the environment. The law also allowed EPA to recover its cleanup costs from the parties responsible for the contamination. Through Cooperative Agreements, EPA authorizes states to provide oversight of cleanups at federal priority sites which must meet specific criteria to be listed on the National Priorities List (NPL). Only sites on the NPL are eligible for federal Superfund financing.

Sites nominated for the NPL have received a minimum score of 28.5 using the Hazard Ranking System (HRS) developed under EPA's direction. However, a score of 28.5 or more does not guarantee EPA will place a site on the NPL. The HRS model is used to assess the actual or potential threat a site presents to public health or the environment. The model is based on a number of factors such as the physical characteristics and quantity of chemicals present, the size of the population exposed, and the impacts to drinking water supplies, surface water, air, direct human exposure and sensitive environmental systems. The intended purpose of the HRS approach is to identify and focus resources on cleanup of the most hazardous sites. Under HRS-I (the original system) many agricultural chemical sites did not score very high because of a bias toward high population centers. Revisions to HRS-I, referred to now as HRS-II, were recently completed as mandated by the Superfund Amendments and Reauthorization Act (SARA) of 1986. These revisions are supposed to more accurately quantify the risk posed by a site. This includes placing greater emphasis on a population which has already been impacted rather than may potentially be impacted.

In the HRS-II Superfund Chemical Data Matrix (April 1991), many common agricultural chemicals have a toxicity factor score of 100 whereas many other chemicals have toxicity factor scores as high as 10,000. As a result, many agricultural chemical sites will likely have HRS-II scores lower than other kinds of sites.

Cleanup under the federal Superfund program is a complicated and expensive process that is governed by stringent criteria established by EPA in the National Contingency Plan (NCP). One set of criteria, called Applicable or Relevant and Appropriate Requirements (ARARS), are state and federal laws which must be followed during the cleanup of Superfund sites. The NCP refers to dozens of potential ARARS such as RCRA and FIFRA and their supporting regulations, which may apply at Superfund sites.

As a result of the various program requirements, cleanup at federal Superfund sites may cost millions of dollars and take years to complete. Cleanup costs increase dramatically when contaminated soil must be managed as a hazardous waste and incinerated, and when alternative drinking water sources must be supplied. Small businesses, such as agricultural chemical dealers, are not likely to have sufficient assets to pay these high costs. The general approach used by states and EPA to investigate and cleanup these sites will likely be the Superfund program. The Minnesota Environmental Response and Liability Act (MERLA) created a state superfund program to manage sites not eligible for federal Superfund money, following essentially the same process used in the federal program. The same limitations that apply to CERCLA apply to the state superfund. Few agricultural chemical sites were traditionally remediated under CERCLA or, in Minnesota, under MERLA because superfund resources were limited and usually were focused on sites near more densely populated areas or on sites containing extremely toxic materials.

Future agricultural chemical site cleanups also may be affected by the 1988 FIFRA amendments. FIFRA regulations will require agricultural chemical dealers to construct pesticide storage containment facilities. Many of these facilities might be built over existing contamination, and future cleanups could require removal of the containment facility. In Minnesota, agricultural chemical dealers are required by state rules to construct pesticide storage containment facilities by June 1992. Many of these dealers have performed soil sampling at their proposed construction sites, and contaminated areas are being cleaned up prior to construction of containment facilities.

The Minnesota Agricultural Chemical Cleanup Fund

The awareness of contamination at agricultural chemical facility sites became apparent in Minnesota in the mid 1980's. The Minnesota Department of

Agriculture (MDA) initiated legislation giving it explicit authority to request, order, or compel responsible parties to cleanup contaminated sites. In 1988, the legislature debated the development of a fund to provide reimbursement to those who properly reported and remediated accidental and incidental spill sites. Passed as a major component in the Minnesota Comprehensive Ground Water Protection Act of 1989, the reimbursement fund was to be administered by the Minnesota Department of Agriculture. In addition, the legislature also gave the MDA expanded "superfund" authority in order to cleanup sites where no responsible party was found or where the responsible party was non-cooperative.

The Agricultural Chemical Response and Reimbursement Account (ACRRA) is intended to reimburse persons for corrective action costs incurred in cleanups after July 1, 1989. The account is funded by annual surcharges assessed to pesticide and fertilizer manufacturers ("registrants"), distributors, applicators, and agricultural chemical dealers. The amount of surcharge levied is largely determined by the current ACRRA balance and projected fund exposure; the account has a statutory required minimum balance of $1 million and a maximum balance of $5 million.

Ordering reimbursement from the ACRRA is the responsibility of a five member Agricultural Chemical Response Compensation Board (ACRRA Board). The industry funding this account has a majority interest in the Board's membership; one member represents farmers, one member represents dealers, and one represents manufacturers. These three citizen members are appointed for four-year staggered terms by the incumbent state governor; the Commissioners of the Minnesota Departments of Agriculture and Commerce complete the membership.

The Board is charged with determining if costs requested for reimbursement by eligible persons are reasonable and necessary. Eligibility is limited by law to responsible persons or owners of real property but does not include local, state or federal units of government. The state statutes allow costs to be reimbursed according to the following formula: 90% of the costs greater than $1000 and up to $100,000, and 100% of costs greater than $100,000 and up to $200,000. The $200,000 ceiling of coverage will effectively mean that expensive site (or emergency) cleanups will be only partially reimbursable with the balance the obligation of the responsible party. The ACRRA statute and the rules promulgated by the Board deny certain types of costs: for example, costs related to the repair or replacement of facility structures or equipment, decreased property values, reimbursement for the eligible person's own time spent in planning and administering the corrective actions, attorney's fees, and costs associated with providing alternative sources of drinking water.

The exclusion of the cost of drinking water may be problematic. If and when drinking water has been contaminated by agricultural chemicals, costs for an alternative source are potentially great. In those instances the responsible party is liable for all costs under state and federal law. Currently, there is no other fund available to share or defray those costs. The state superfund may be utilized to provide water in cases of emergency; however, costs incurred by the state must be recovered by the state upon determination of responsible party.

The ACRRA Board may reduce requested reimbursements if the Board determines that a violation of applicable Minnesota agricultural chemical law and regulations caused the incident. Because the Board makes the determination, independent from enforcement by the Department of Agriculture, it has the ability to develop criteria and rules on which to make reductions.

Conclusions

The scope of the problem resulting from the past thirty years of accidental and incidental spillage of agricultural chemicals at mixing, loading and handling sites is large. Discussion at the federal and national level, coupled with the initiation of remediation efforts by some states, has uncovered another alarming problem: very little research has been conducted on this issue. Accordingly, alternatives to land spreading are expensive and untested. At this time, insitu assessment and remediation techniques are limited and costly. Cost containment, which is necessary for small businesses to remain viable, is difficult, but essential.

The lack of research on assessment and remediation of agricultural chemical spill sites reflects the general lack of basic research on regulatory issues related to agriculture. In 1990, an informal group of industry, state and federal representatives known as the Agricultural Remediation Technology Consortium (ARTC) met to discuss the issue of agricultural spill site remediation. The lack of research stimulated the development by state agencies of draft research proposals that were discussed by ARTC. In addition to the need for research, the need for a reassessment of federal statute interpretation was recognized. Federal statutes were not designed for the agricultural chemical spill site situations; although the 1988 FIFRA amendments address the issue, it is unclear at this time how FIFRA will interact with RCRA. With a few exceptions, states are only now beginning to address the issue.

Currently, legislation much like the Minnesota ACRRA statute is under debate in Iowa *(11)*, and a modified version for catastrophic events is in place in Illinois. Nonetheless, it is evident that a significant environmental problem exists, and solutions need to be developed to address this serious concern. States, some of which are actively addressing the issue, need support through research, funds, and legislation at both the state and national level.

Literature Cited

(1) Zuelsdorf, N. Wisconsin Department of Agriculture, Trade and Consumer Protection, Madison, Wisconsin. Personal communication. August 22, 1991.

(2) Holden, P.W. 1986. Pesticides and Groundwater Quality. National Academy Press, Washington, D.C. 124 pp.

(3) Klaseus, T.G., Buzicky, G.C. and Schneider, E.C. Pesticides and Groundwater: A Survey of Selected Minnesota Wells. Minnesota Department of Health and Minnesota Department of Agriculture. February, 1988.

(4) Victor, W.R. Impact on Ground Water by Distributors of Agricultural Chemicals. Proceeding of Agricultural Impacts on Ground Water - A Conference. National Water Well Association. Omaha, Nebraska. August 11-13, 1986.
(5) Hallberg, G.R. Overview of Agricultural Chemicals in Ground Water. Proceeding of Agricultural Impacts on Ground Water - A Conference. National Water Well Association. Omaha, Nebraska. August 11-13, 1986.
(6) Habecker, M.A. 1989. Environmental contamination at Wisconsin pesticide mixing/loading facilities; case study, investigation and remedial action evaluation. Wisconsin Department of Agriculture publication. 80 pp.
(7) RCRA Superfund Hotline 1-800-424-9346.
(8) EPA Overview of RCRA Land Disposal Restrictions (LDRs), Directive 9347.3-01FS. July, 1989.
(9) EPA Complying With the California List Restrictions Under Land Disposal Restrictions (LDRs), Directive 9347.3-02FS. July, 1989.
(10) EPA Overview of RCRA Land Disposal Restrictions (LDRs), Directive 9347.3-01FS. July, 1989.
(11) Frieberg, D. Iowa Fertilizer and Chemical Association, Des Moines, Iowa. Environmental Cleanup of Fertilizer and Ag Chemical Dealer Sites. March 7, 1991.

RECEIVED July 15, 1992

Chapter 22

Experimental Design for Testing Landfarming of Pesticide-Contaminated Soil Excavated from Agrochemical Facilities

Allan S. Felsot[1], J. Kent Mitchell[2], T. J. Bicki[3], and J. F. Frank[4]

[1]Illinois Natural History Survey, 607 East Peabody Drive, Champaign, IL 61820
[2]Department of Agricultural Engineering and [3]Department of Agronomy, University of Illinois, Urbana, IL 61801
[4]Andrews Environmental Engineering, Inc., Springfield, IL 62707

Of the various methods proposed to remediate pesticide-contaminated soils, landfarming may be the closest to immediate implementation. The technique involves the spreading of waste-contaminated soil on agricultural or noncropped land to stimulate degradation, transformation, and/or immobilization of contaminants. It is particularly suited for pesticide-contaminated soils because it is a technology already used to dispose of municipal wastewater and sludges, and the pesticides involved are usually registered by the U.S. EPA. This report describes an experimental design for testing the feasibility and safety of landfarming soil contaminated with alachlor, trifluralin, atrazine, and metolachlor. The objective of the design was to compare herbicide behavior following landfarming of different rates of contaminated soil with herbicide behavior following conventional application by spraying at the same rates. A manure spreader was used to apply the contaminated soil. The sampling protocol was designed to minimize coefficients of variation for the mean residues. Field plots were installed with runoff collectors and salt-water samplers to measure translocation of the herbicide contaminants. Toxicity of landfarmed herbicide residues to crops, weeds, and algae were determined by appropriate assays. A preliminary analysis of herbicide residues in soil immediately after application showed that coefficients of variation were lowered to acceptable levels by collecting six individual cores per 30 m^2 plot. Target rates of application, however, were only reached for landfarmed treatments and were two- to three-fold lower than expected for sprayed treatments. Concentrations of herbicides in runoff water increased with initial rate of application and were greater from landfarmed plots than from sprayed plots. Total herbicide loads in a cumulative three-day runoff event did not generally differ among treatments. Inhibition of photosynthesis after exposure of algae to runoff water failed to follow a clear dose-response relationship. Toxicity to soybeans occurred at the lowest rates of application, and biological activity of landfarmed herbicide residues was clearly indicated by inhibition of weed growth.

0097–6156/92/0510–0244$06.00/0

Soils at agrichemical retail facilities throughout the United States are frequently contaminated by pesticides during loading and rinsing operations, even after containment facilities have been installed (*1*). In some cases, historical standard operating practices have left a legacy of high pesticide concentrations in soil and ground water that have affected nearby residential water supplies (*2*). In some cases, contamination is serious enough to warrant a cleanup order from a state or federal regulatory agency; in other cases, the cleanup may be voluntary as a result of the desire to install new structures or to transfer the property. In either case, the soil must somehow be remediated, but the technical options may be limited and prohibitively expensive.

Current options for cleanup of soil include in situ treatment or excavation followed by treatment (displacement) (*3,4*). In situ methods like vitrification and bioremediation are not presently practical nor commercially available for pesticide-contaminated soils. Currently used displacement techniques include landfilling and incineration. Unfortunately landfilling is becoming a less viable option because of space limitations, and incineration of large quantities of soil is too expensive and not easily accessible for small agrichemical facilities. Treatment of excavated soil by bioremediation may be more suitable to rural facilities because it is an easily transported technology that is cheaper than incineration.

Bioremediation methods under research and development that seem suitable for soil cleanup include composting, bioreactors, and landfarming. Of these three methods, landfarming may be the closest to immediate implementation. Landfarming, also known as land treatment, land application, or land spreading, is a "managed treatment and ultimate disposal process that involves the controlled application of a waste to a soil or soil-vegetation system (*5*)." The waste is spread on agricultural or noncropped land to stimulate degradation, transformation, and/or immobilization of contaminants. The technique is particularly suitable for pesticide-contaminated soils because the practice involves a technology that has been used for many years to dispose of municipal wastewater, sludges, and petroleum refinery wastes (*6*); furthermore, the pesticides in contaminated soils are usually registered by the U.S. EPA for application to soil.

Because of the need for easily implemented methods to dispose of pesticide-contaminated soil during cleanups of agrichemical facilities and the relative ease of excavating and spreading soil, the Illinois Legislature recently authorized through 1992 the Illinois Department of Agriculture (IDOA) to permit the land application of pesticide-contaminated soil at agronomic rates (*7*). The authorization allowed IDOA to prescribe operational control practices to protect the site of application. Successful application of landfarming, however, relies on a detailed site assessment to accurately prescribe the nature and extent of the waste, establishment of cleanup objectives, and development of a remedial action strategy (*8*). Although many states are considering regulations allowing landfarming for disposal of pesticide-contaminated soils, the effectiveness of the technology, methods for stimulation of degradation rates, and possible off-site movement of contaminants has hardly been studied.

Remediation of Pesticide Waste in Composting and Landfarming Systems

Most research efforts have focused on bioremediation of chlorophenolic wastes in soil that arise from wood preserving operations (e.g., *9-10*). Soils previously exposed to pentachlorophenol (PCP) have exhibited enhanced rates of biodegradation when retreated, suggesting microbial adaptation and utilization of the pesticide as an energy source. A combination of bioaugmentation and biostimulation has been used in combination with composting in windrows to treat chlorophenol-contaminated soils in Finland (*11*). The effects of tilling and fertilization on biodegradation of PCP were studied in "landfarming chambers" (*12*); degradation rates of PCP were too slow to meet acceptable treatment standards after 90 days of incubation.

Liquid pesticide wastes have been successfully detoxified by pretreatment prior to soil disposal. Chlorinated hydrocarbon pesticides biodegraded during sewage treatment, and diazinon, parathion, and dieldrin degraded rapidly when composted with cannery wastes (*13*). Liquid wastes of alachlor and atrazine have been exposed to UV light and ozonation prior to disposal in soil (*14,15*). These pesticides are cometabolically degraded, and the pretreatment degrades the compounds into products mineralized by soil microorganisms.

Land application of composted cotton gin wastes containing pesticides resulted in very low to undetected residues after incorporation of the waste into the soil (*16*). Pesticide waste taken from a highly contaminated soil evaporation pit in California was partially detoxified by amendment with a variety of inorganic and organic nutrients and by holding it under aerobic or anaerobic conditions (*17*). Corn plant residue, corn meal, and municipal sewage sludge effectively stimulated degradation of high concentrations of alachlor in soil (*18*). A ten-fold dilution of alachlor and metolachlor-contaminated soil with uncontaminated soil to simulate landfarming also stimulated pesticide degradation (*19*). Degradation of pesticides adsorbed on lignocellulosic materials was enhanced by mixing into peat-based composting bioreactors (*20*).

Soil contaminated with the herbicides alachlor, metolachlor, atrazine, and trifluralin were excavated from an agrochemical facility in Piatt Co., IL and land applied to corn and soybean fields (*21,22*). Alachlor and metolachlor were more persistent when applied in the contaminated soil than when applied as fresh herbicide sprays (*23*). Minor phytotoxicity to soybeans treated with the highest rates of waste soil occurred in the field and in greenhouse assays. The mixture of corn and soybean herbicides present in the soil was viewed as a potentially limiting factor in attempting to landfarm the pesticide wastes. Unresolved questions included the translocation of the pesticides and the mechanism of prolonged persistence in the contaminated soils.

Testing the Feasibility and Safety of Landfarming Pesticide-Contaminated Soil

In the Corn Belt, candidate land to receive contaminated soils will likely be fields that cannot be taken out of production. Unresolved questions from previous landfarming experiments included the translocation of pesticides from the receiving land, prolonged persistence of pesticides in contaminated soils, and potential for crop phytotoxicity (*21,22*). Because landfarming is a leading candidate for treatment of pesticide-contaminated soils, these questions must be accurately assessed to help define appropriate guidelines and regulations. This paper describes the design of a field experiment to test the degradation, translocation, and phytotoxicity of pesticide contaminants in landfarmed soil. In addition to describing the design and sampling procedures, preliminary results of soil residues and runoff monitoring are presented.

Hypothesis. Appropriate experimental design depends on a well defined hypothesis. In this case, the hypothesis is stated as criteria of feasibility that after completion of the experiments should enable a judgement about the effectiveness and safety of landfarming. Thus, successful remediation of pesticide-contaminated soil is determined by the satisfaction of four criteria:

(1) Pesticide residues in control (untreated plots) and contaminated soil-treated plots (hereafter referred to as landfarmed plots) after two growing seasons do not differ significantly. In the current design, pesticide behavior in landfarmed plots is compared to pesticide behavior in freshly sprayed plots, which serve as positive controls.
(2) Contaminant concentrations and toxicity in leachates and runoff water are at levels not significantly different from or are even lower than concentrations and toxicity in the controls;
(3) Crop phytotoxicity is not greater than expected from conventional sprays;
(4) Residues of pesticides in crops should not violate established U.S. EPA tolerances.

Experimental Design. The experiment consisted of three main effects treatments arranged in a completely randomized design. The treatments were herbicide-contaminated waste soil (landfarmed plots), herbicide sprays (freshly sprayed plots), and no pesticide application (checks). Contaminated soil and herbicide sprays were applied at three rates of application based on the most prevalent pesticide in the waste soil. Each combination of pesticide treatment and rate was replicated four times. The checks were replicated six times; three of the checks were hand-weeded, and three were left unweeded.

Plot Design. Replicates of the pesticide and no-pesticide treatments were randomly assigned to one of 30 experimental plots at the University of Illinois (UI) Cruse Farm. The field encompassed about 0.6 ha on a 3-5% sloped gradient. The soil was classified as a Catlin silt loam (Fine-silty, mixed, mesic Typic Argiudoll). Each plot comprised an area 10 m long x 3 m wide with the length of the plot oriented up-and-down the slope (Figure 1).

The down-gradient end of each of two plots was fitted with metal troughs that directed surface runoff into a 208-L polyethylene barrel housed in a 2.4 m x 2.4 m x 1.2 m pit. A second 208-L barrel sat a a slightly lower elevation to the primary collection barrel and received runoff overflow through a 1:9 flow splitter. The system was designed to receive all the runoff from a 15-year frequency storm. Surface runoff from individual plots was contained by delineating each plot with soil berms along the length and by a metal barrier along the up-gradient end.

Two porous-cup soil-water samplers were placed along the midline of each plot at a distance of 3.7 m from each end. The samplers were installed at a 45° angle to the horizontal so that the ceramic sampling cup was at a perpendicular distance of 60 cm from the surface. Pressure-vacuum sampling tubing issuing from the samplers was buried 40 cm below the soil suface and extended beyond the soil berm.

Herbicide-Contaminated Soil. During April 1990, water used to fight a fire at a pesticide warehouse in Lexington, IL flooded the soil surrounding the building and deposited high concentrations of trifluralin and lesser concentrations of atrazine, alachlor, and metolachlor. The soil was excavated and stored until August 1990 at a farm where it was eventually disposed of by landfarming (*7,8*). Analysis of 18 individual cores (5 cm diam. x 10 cm deep) just prior to landfarming showed that trifluralin was the primary constituent; the average concentration was 158 ± 247 ppm (range 3-1003 ppm) (*8*). Approximately 20 Mg of this waste soil was transported to the UI Cruse Farm during August. The soil was covered with a black plastic sheet and stored through the winter. During March 1991, six cores (5 cm diam. x 10 cm deep) were collected randomly from the pile. The following concentrations were found: 118 ± 58 ppm trifluralin, 18 ± 14 ppm metolachlor, 1 ± 1 ppm atrazine, and 1 ± 1 ppm alachlor.

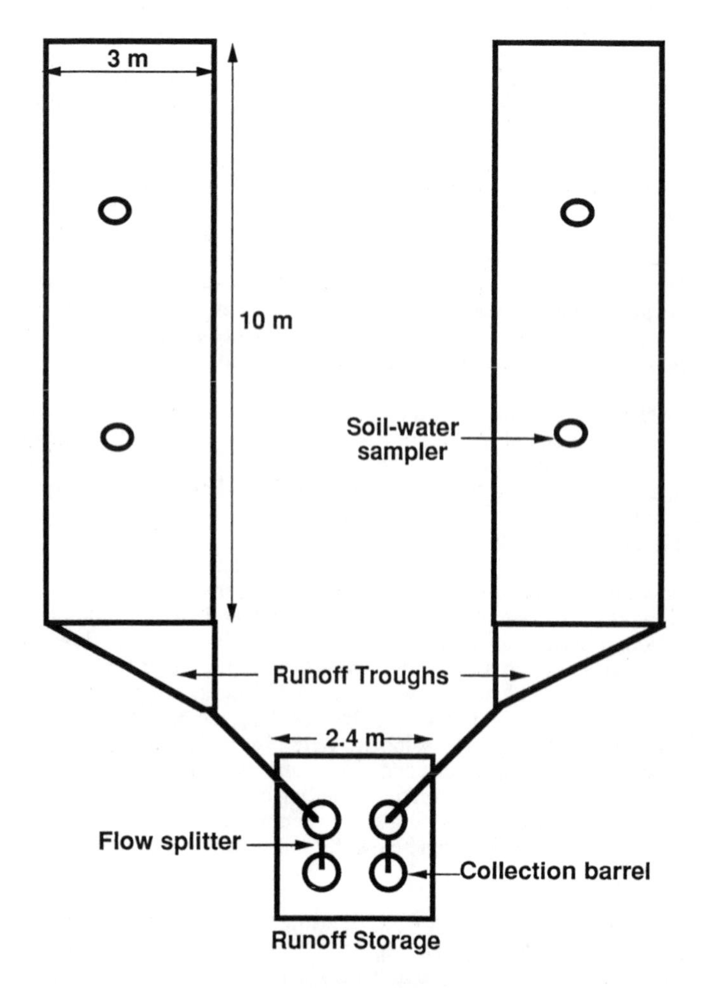

Figure 1. Plot design for landfarming experiments at the University of Illinois. Thirty plots were constructed and received either pesticide-contaminated soil or fresh herbicide sprays at different rates of application.

Because we were interested in landfarming high concentrations of a greater diversity of herbicides, a simulated spill of alachlor was created on 7 April 1991 by pouring 9.5-L of Lasso 4E (480 g alachlor/L) in 30-cm deep trenches that were dug into the surface of the pile; the trenches were then filled with soil. On 25 April 1991, 1.4 L of Aatrex 4F (480 g atrazine/L) were spilled on the surface of the pile, and the soil was overturned with a spade. On 26 April 1991, a front loader mixed the pile of contaminated soil by completely overturning it in one direction and then overturing it a second time in a direction perpendicular to the first. The soil was then piled about 0.6-0.9-m high within a 7.6 m x 3.0 m area. On 31 May, 10 cores (5 cm diam. x 15 cm deep) were collected along a diagonal transect laid across the surface of the pile, and 10 cores were also collected from a depth of 30-45 cm. Herbicide concentrations (oven-dry weight basis) determined in individual cores averaged 172 ± 99 ppm alachlor, 99 ± 53 ppm trifluralin, 18 ± 9 ppm metolachlor, and 14 ± 11 ppm atrazine.

Rates of Application: Target rates of application were based on the concentration of alachlor in the contaminated soil pile. The maximum legal application rate of alachlor is 4.48 kg ai/ha; this rate represented the 1X level. Five and 10 times the maximum rate were also determined. The theoretical rates of application of trifluralin, metolachlor, and atrazine were based on their soil concentrations in proportion to the concentration of alachlor at the 1X, 5X, and 10X levels (Table I).

Table I. Theoretical Application Rates (kg ai/ha) of Herbicide Contaminants

Proportional Rate	Alachlor	Trifluralin	Metolachlor	Atrazine
1X	4.5	2.6	0.5	0.4
5X	22.4	12.8	2.4	1.8
10X	44.8	25.7	4.8	3.6

Application of Contaminated Soil and Sprays. The weight of soil needed per plot to give an equivalent alachlor application rate of 4.5, 22.4, and 44.8 kg ai/ha was based on the average concentration (172 ppm) determined in May. Thus, on an oven-dry weight equivalent basis, 78, 391, and 782 kg of contaminated soil were needed to produce proportional application rates of 1X, 5X, and 10X. On 6 June 1991 the appropriate amount of soil was loaded into a 1.3-m wide manure spreader from a front-end loader that had been calibrated by weighing empty and full. The comparatively small plot size necessitated delivery of the soil from the manure spreader without the use of the beater blades; the soil dropped off the back of the spreader as the drive chains moved. The soil was applied in one pass and then raked evenly across the entire 3-m width of the plot.

For the freshly sprayed treatments, enough alachlor (Lasso 4E, 480 g/L), trifluralin (Treflan EC, 480 g/L), metolachlor (Dual 4E, 960 g/L), and atrazine (Aatrex 4L, 480 g/L) were mixed together with tap water to give theoretical application rates equivalent to the proportional rates calculated for the landfarmed plots (Table I). On 5 June 1991 the spray was delivered from a tractor-mounted boom calibrated to deliver 336 L of spray per ha.

Plot Preparation and Planting. After application of contaminated soil and herbicide sprays, all plots were disked to a depth of 10 cm in an up-and-down slope direction; the soil surface was smoothed with a rolling bar cultivator. On 7 June 1991 four rows of soybeans (*Glycine max* L.) were planted in each plot.

Sampling of Applied Soil and Sprays. During application of contaminated soil and herbicide sprays, three aluminum pans (29.2 x 19.3 cm) were placed along the vertical midline of each plot to intercept the applied material. The pans were packaged in individual polyethylene bags and returned to the laboratory for analysis. Contaminated soil intercepted on the landfarmed plots was weighed, sieved through an 8-mesh screen (3-mm openings), and stored at -20°C before analysis. A subsample was oven-dried to determine the percentage moisture content. Bags containing the pans with intercepted spray material were frozen immediately.

Immediately after application, six cores (5 cm diam. x 10 cm deep) were randomly collected from each plot (landfarmed and sprayed). The cores were placed in individual polyethylene bags and returned to the laboratory for analysis. Each core was sieved and stored at -20°C.

Water Sampling. Prior to sampling runoff, the depth of water to the bottom of the barrel was measured within 24-48 hours after a runoff event to determine the volume of runoff. The water and sediment were stirred with a paddle, and water was collected for pesticide analysis by submersing two 500-mL glass bottles into the barrel. An additional two samples of mixed sediment and water were collected by submersing a Nalgene bottle into the barrel; this sample was used to determine sediment concentrations necessary to calculate the weight of eroded soil. Samples for pesticide analysis were returned to the laboratory shortly after collection and stored at 4°C for 48-72 hours before analysis. Prior to extraction, water was separated from sediment by filtration through a glass microfiber filter (Whatman no. 934 AH); sediment was weighed wet and after air-drying overnight (*24*) Samples were then frozen at -20°C.

Soil-water was collected within 3-4 days of a significant rainfall event or at least once a month. Soil-samplers were kept under a negative pressure of 414 kPa. During sampling, the pressure was released and water was pumped under positive pressure into a glass bottle. Water from the two samplers in each plot were composited and stored at 4°C for about 2-3 days before analysis.

Toxicity Assays. Phytotoxicity of herbicide residues in soil was determined by counting total emerged soybean plants in each plot on 5 July 1991. Total number of weeds in each plot were counted on 19 July 1991. Toxicity of herbicide residues in runoff water were determined by the algal photosynthetic inhibition bioassay (*25*).

Analytical Methods. Pans containing intercepted spray residues were extracted by rinsing with acetone. Plastic bags were rinsed twice with hexane and then cut into small pieces and stirred with 150 ml of hexane for 30 min. Extracts were diluted to a final volume of 250 mL before analysis.

Thirty-grams of soil were slurried with 12 mL of glass-distilled water and extracted three times by stirring with 60 mL glass-distilled ethyl acetate. The solvent was decanted into a standard taper flask and rotary evaporated to dryness under vacuum at 35°C. The extract was reconstituted in 10 mL of ethyl acetate.

Water (usually 500 mL) was extracted twice in a separatory funnel with 50 mL of glass-distilled methylene chloride. The methylene chloride extract was passed through oven-dried (110°C) sodium sulfate and then rotary evaporated like the soil extracts. The extract was reconstituted in 1 mL of ethyl acetate.

Herbicide residues were determined on a Packard model 438 gas-liquid chromatograph with a nitrogen-phosphorous detector and autosampler. The column (90 cm 5% Apiezon + 0.1% DEGS) and operating conditions have been described previously (*26*). The identity of the residues were qualified by comparison to the retention time of external standards that were injected after every 7 extract samples. Residues were quantified by multilevel calibration with external standards.

Limits of detection were 150 ppb in soil and 1 ppb in water. Extraction efficiencies from soil and water were determined by use of fortified blanks, which consisted of soil and runoff water collected from untreated plots. Recovery of all herbicides from soil were greater than 80% (*21*). From water fortified at 1 ppb, recoveries were 64% for trifluralin and >85% for atrazine, alachlor, and metolachlor. Data were not corrected for extraction efficiencies.

Results and Discussion. Our previous study (*21,22*) indicated that landfarming required at least two years to remediate soil containing elevated concentrations of alachlor and metolachlor; concentrations of these contaminants from a theoretical 1X application rate were not significantly different after two years than concentrations of herbicides in untreated soil. Thus, the present landfarming experiment was designed to last a minimum of two years. The objective of this report was to assess the efficiency of the experimental design with respect to initial recovery of pesticide residues, sampling variability, runoff collection, and toxicity assays. Thus, only soil residues immediately after application are presented; also, the results from a runoff event in July 1991 are shown to illustrate how runoff data is analyzed and compared.

Initial Herbicide Residues. Because a major criterion of feasibility of landfarming was to be determined by comparing herbicide residue behavior in landfarmed and sprayed plots, an accurate assessment of actual material applied and initial concentrations in soil was imperative. The actual rates of herbicides applied were determined by analysis of spray and soil intercepted by pans placed in three locations in the plots and compared to the target rates of application (Table II). In

Table II. Targeted and Calculated Rates of Application of Herbicides

	Rate of Application (kg/ha) Calculated from Recovered Residues			
Targeted	Herbicide Spray		Landfarmed Soil	
Rate	Interceptor Pans	Soil Cores	Interceptor Pans	Soil Cores
Alachlor				
4.5	2.3	1.9	4.8	6.4
22.4	13.9	11.9	32.8	32.5
44.8	20.2	17.3	48.4	54.7
Trifluralin				
2.6	0.9	0.8	3.0	4.1
12.8	7.1	6.1	20.8	19.0
25.7	10.8	9.3	30.7	30.3
Metolachlor				
0.5	0.2	0.2	0.5	0.7
2.4	1.1	1.0	3.3	3.6
4.8	1.6	1.3	4.9	6.4
Atrazine				
0.4	0.2	0.1	0.4	0.5
1.8	1.0	0.8	2.7	3.6
3.6	1.7	1.5	4.0	6.2

each case, residues intercepted by pans in sprayed plots were approximately 2-3 times lower than the theoretical application rate. Average residues recovered from soil cores yielded rates of application similar to those calculated from interceptor pans and confirmed the lower than expected application rate of the sprays. An analysis of the actual spray solution indicated that the amount of pesticides added to the tanks was near theoretical amounts; thus, it was concluded tht the application of finished spray at a targeted rate of 336 L/ha was not achieved. Indeed, an excessive volume of spray solution remaining after application supported this hypothesis and suggested that either the sprayer was improperly calibrated and/or the tractor speed was excessive.

After application of contaminated soil, soil collected in the interceptor pans was weighed and scaled up to a plot area basis (Table III). For the 1X and 10X treatment, the actual weight of soil applied per plot was within 8% of the target rate; the weight of soil applied in the 5X plot was 26% greater than the target weight. In contrast to residues recovered after spraying, residues recovered after application of contaminated soil tended to be higher than the expected amounts (Table II). The largest discrepancies (approximately 150-160% of target rates) were observed in the 5X plots as expected from the weights of soil actually applied.

Table III. Dry Weight (kg) of Contaminated Soil Applied Per Plot

Proportional Rate	Targeted Weight	Actual Weight
1X	78	72
5X	391	493
10X	782	729

Because average initial residues of the herbicides in sprayed plots were significantly below the average residues in the landfarmed plots, a direct comparison of degradation is feasible only if the kinetics of degradation are first order, i.e., the rate of degradation is independent of the initial concentration. If so, then residue data could be normalized to a percentage of actual amounts applied, and first-order kinetic constants calculated for sprayed and landfarmed degradation curves could be compared directly. Similarity of degradation kinetics can be tested under laboratory conditions using different starting concentrations of freshly applied herbicide. Whatever the result of such an experiment, an appropriate kinetic model must be employed to directly compare the degradation rate of herbicides between the two kinds of treatments.

Intra- and Inter-plot Variability of Herbicide Residues. In a previous landfarming study, comparisons between sprayed and landfarmed treatments were complicated by large coefficients of variation, especially with samples collected immediately after application (*21*). Resolution of differences between treatments by analysis of variance was only observed after a full year of sampling. In an attempt to lessen sampling variance, the number of cores taken per plot in the current experiment was increased to six; furthermore, each core was analyzed individually. Concentrations in the six cores were averaged before an analysis of variance was conducted on the four true experimental units or replicates. An analysis of the alachlor and trifluralin concentrations in the 1X and 10X treatments showed that the standard deviations for average concentrations calculated from individual soil cores within a plot were always larger than the standard deviations for average concentrations calculated across the four replicate plots (Figure 2,3). As suggested by Figures 2 and 3, the average coefficient of

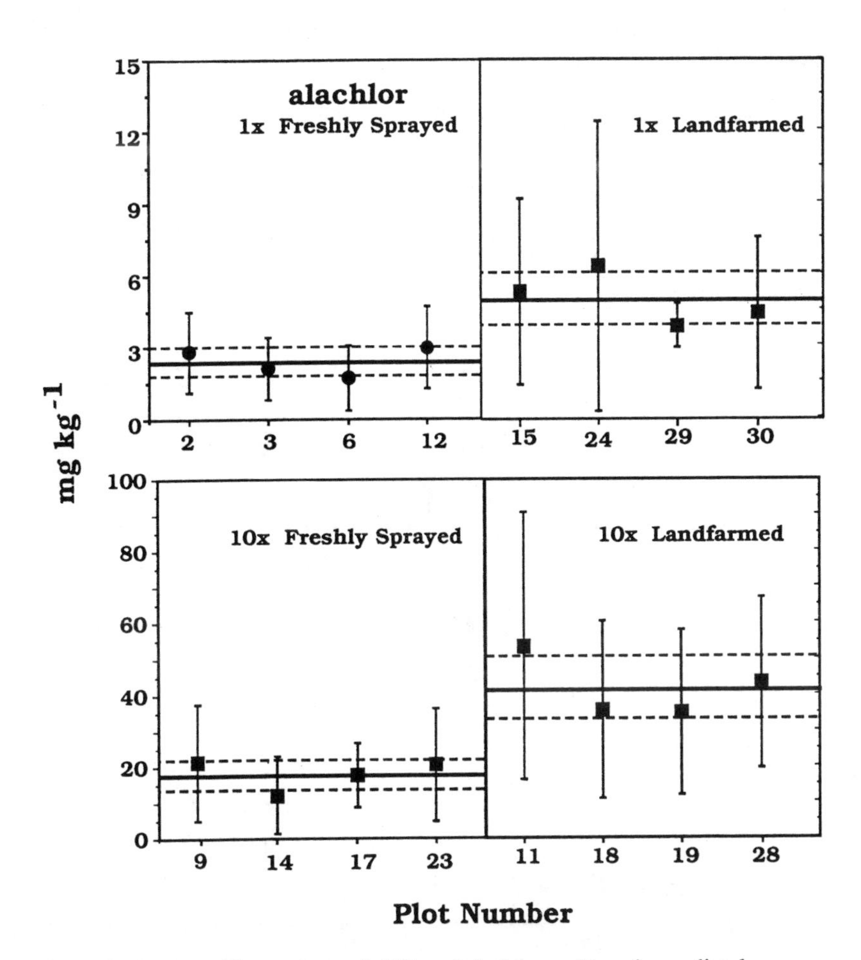

Figure 2. Intra- and inter-plot variability of alachlor residues immediately following application by landfarming or spraying. Plot numbers represent replicate experimental units. Vertical lines represent standard deviations about the mean of six individual soil cores from within one plot. Heavy horizontal line represents the mean of four replicate plots; broken horizontal lines represent the upper and lower limit of the standard.

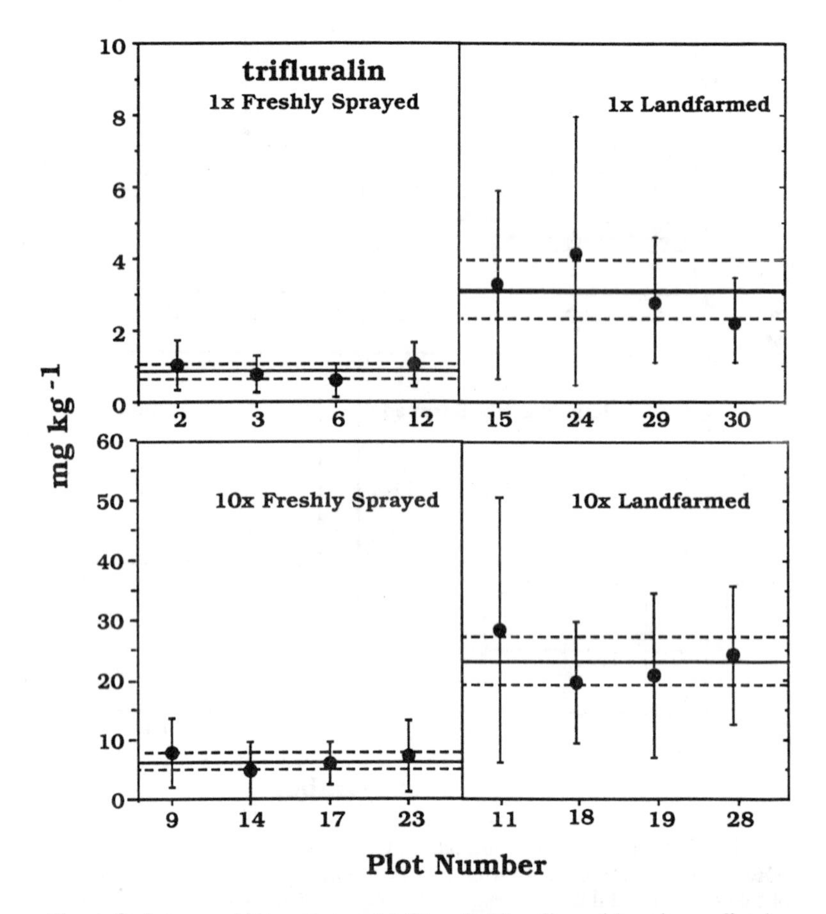

Figure 3. Intra- and inter-plot variability of trifluralin residues immediately following application by landfarming or spraying. See Figure 2 for further explanations.

variation for all herbicide concentrations determined in individual soil cores within a plot (71.7 ± 23.7%) were much higher than the average coefficients of variation determined by averaging concentrations between plots (27.2 ± 9.6%). Thus, spatial variability in herbicide concentrations for similar treatments across the whole field seemed to have been lowered by averaging out the variability within a plot.

Studies directed specifically at the nature of variability associated with sampling pesticide residues in soil have shown that coefficients of variation can be lowered to about 20% at best (*27,28*); such reductions are only achieved by increasing the number of samples per replicate experimental unit. Although more expensive, collecting a large number of subsamples per replicate allows better resolution of differences between treatments if such differences exist. When testing waste disposal methodologies, such resolution is desirable to promote confidence in environmental safety of the process or to discern the most efficacious process from among many possible techniques. In the case of waste disposal by landfarming, intensive sampling of waste-contaminated soil would allow a better assessment of initial loading rates. Furthermore, actual field residues following application could vary by as much as two-fold from the mean (*28*); such variance could have important implications for crop phytotoxicity. Accurate assessment of sampling variance would therefore be important for predicting environmental safety during landfarming.

Concentration of Pesticides in Runoff Water. Past studies have shown that herbicides like alachlor are largely transported in surface runoff rather than in eroded soil (24), so only the concentrations of herbicides collected in the water are discussed. Precipitation totalled 1.6 cm during the month following application of the herbicides. A cumulative rainfall of 5.9 cm during July 10-12 produced three separate runoff events. To compare the effect of spraying and landfarming on herbicide runoff, herbicide loads in individual runoff events were calculated from the product of total runoff volume and concentration; the data were cumulated and expressed as a percentage of the initially applied amounts recovered in the aluminum interceptor pans.

Volumes of runoff from 1X, 5X, and 10X sprayed plots averaged 202, 154, and 207 L, respectively; from 1X, 5X, and 10X landfarmed plots, volumes of runoff averaged 103, 127, and 74 L, respectively. Concentrations of herbicides increased as the rate of application increased (Table IV); also, concentrations in runoff were higher from landfarmed plots than from sprayed plots. Although runoff volumes from landfarmed plots were smaller than runoff volumes from sprayed plots, differences in concentrations were expected because residues were 2-3 times greater in the landfarmed plots than in the sprayed plots. As a percentage of applied amounts, the load of alachlor and atrazine recovered in runoff from landfarmed plots was significantly greater ($p<0.05$) than the load from sprayed plots at the 1X rate only (Figure 4). Loads of herbicides from all other treatments were not statistically different (Figure 4). Previous studies with atrazine have also shown increasing loads in runoff as the rate of application increased (*29*); however, runoff losses as a percentage of amount applied were not proportional to application rate.

Atrazine load as a proportion of the applied amount was significantly greater than the loads of all other pesticides. These data are consistent with the relative ubiquity of atrazine in many surface water monitoring studies (*30,31*).

Concentration of Pesticides in Soil-Water. Soil-water was sampled on 12 July 1991. As a result of low precipitation during the first month of the experiment, pesticide leaching was negligible. Atrazine and alachlor were

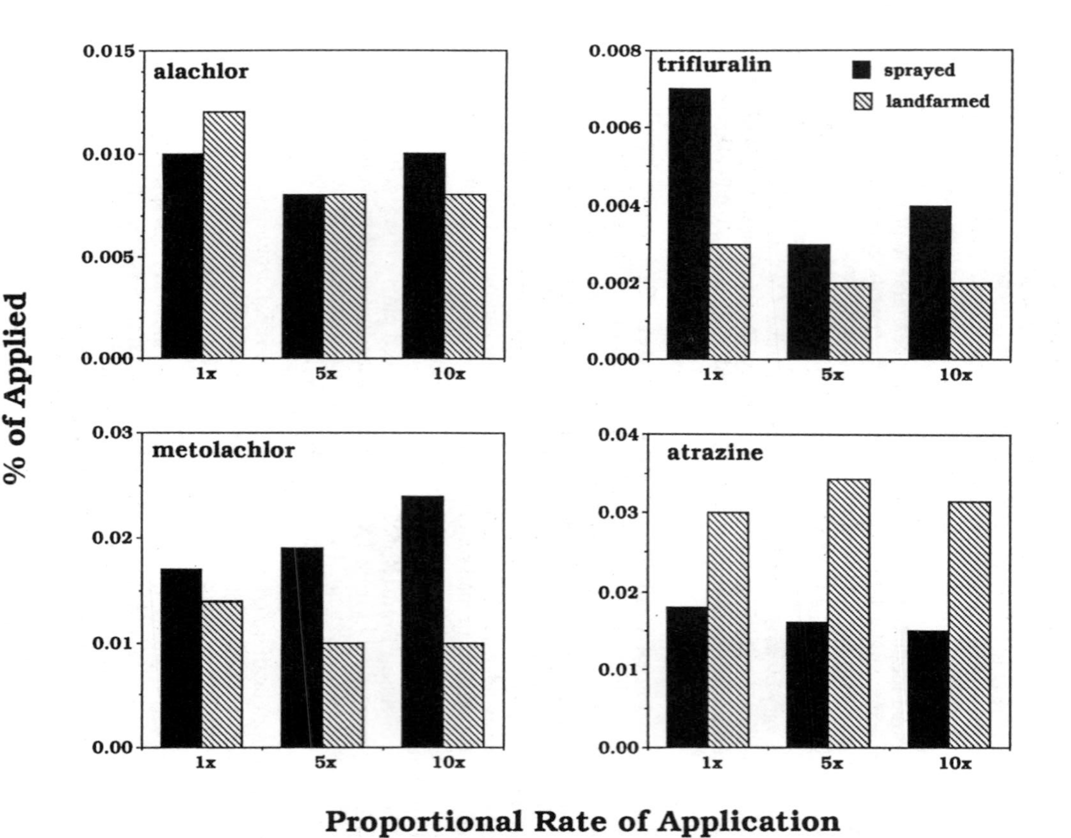

Figure 4. Herbicide load (as a percentage of intial application rate) in a cumulative runoff event during July 1991.

occasionally detected in water from the checks (range of 1.3-23.6 ppb) or 5X landfarmed treatments (range of 4.4-9.1 ppb). Trifluralin and metolachlor were not detected in any of the leachates. Because the field had been treated with a variety of herbicides during previous growing seasons, trace residues in runoff or soil-water were expected.

Table IV. Average Concentration of Herbicides in Runoff Water Collected During July 1991

Proportional Rate of Application	Concentration (ppb)	
	Freshly Sprayed Plots	Landfarmed Plots
Alachlor		
1X	3.6	18.3
5X	24.9	69.3
10X	31.2	176.7
Trifluralin		
1X	1.0	2.8
5X	4.9	10.2
10X	7.6	26.3
Metolachlor		
1X	0.6	2.2
5X	4.4	8.0
10X	6.0	21.5
Atrazine		
1X	0.6	3.8
5X	3.4	23.6
10X	3.9	55.4

Results of Toxicity Assays. Total number of soybean plants in 1X landfarmed plots was significantly less ($p<0.05$) than the number of plants in check plots (Table V). The number of soybean plants in 1X sprayed plots was also lower but not significantly. Phytotoxicity was attributed to the presence of atrazine which is not registered for use on soybeans. The greater inhibition of soybean growth in the 1X landfarmed plots than in the sprayed plots probably resulted from the greater initial concentrations of atrazine (0.39 vs. 0.17 ppm soil). Inhibition of soybean germination was greatest for 5X and 10X plots, which did not differ from one another. Activity against weeds was evidenced by the significantly lower plant numbers in all treatments (except 1X landfarmed plots) when compared to the check.

Inhibition of algal photosynthesis was tested using runoff water from July 11 and 12. Because there was no significant date by response interaction, the data from each day were combined and transformed to a percentage of the response in the check. Algal photosynthesis was significantly inhibited ($p<0.025$) by the 1X landfarming treatment (Figure 5); although not significant, inhibition by water from the 10X landfarmed treatment was also greater than inhibition by water from the sprayed plots. Such results were expected as a result of the higher concentrations of herbicides in the landfarmed plots; however, a normal dose-response relationship was not observed because negligible inhibition of photosynthesis occurred with water from the 5X treatments. Furthermore, the level of inhibition by water from the 10X plots was not significantly different than the level in the 1X

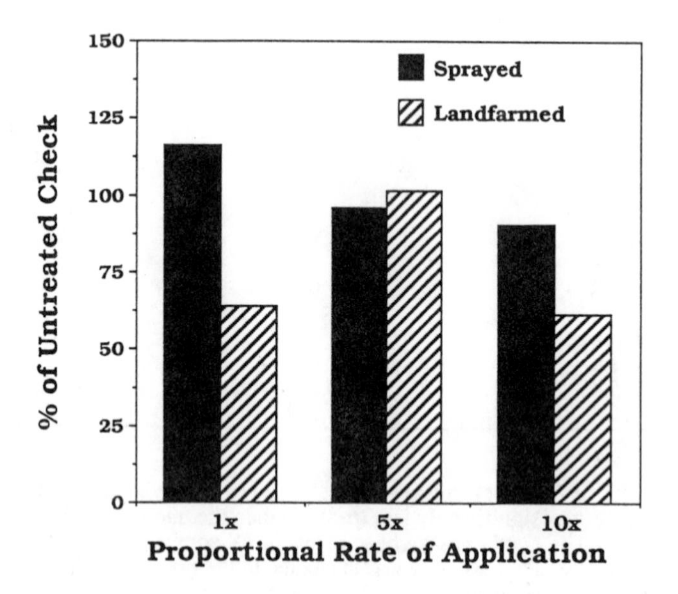

Figure 5. Inhibition of algal photosynthesis in runoff water as a percentage of the untreated check.

plots, although residues in water from the 10X plots were 10-fold greater. In water from the 10X landfarmed plots the highest concentration of atrazine was 36 ppb. In some studies, atrazine, which is known to inhibit photosynthesis, produced adverse chronic effects on phytoplankton populations at concentrations as low as 15 ppb (*32*), but concentrations this low have not been shown to elicit acute effects as measured in the photosynthetic inhibition bioassay. All of these observations suggested that the photosynthetic response to herbicide contamination was highly variable and not necessarily related to the levels in the water.

Table V. Phytotoxicity of Sprayed Herbicides and Landfarmed Soil

Application Rate	No. of Soybean Plants Per Plot		No. of Weeds Per Plot	
	Sprayed	Landfarmed	Sprayed	Landfarmed
Check	459	--	210	--
1X	389	267	90	130
5X	107	140	14	16
10X	92	84	16	18
LSD 1/	114	--	86	--

1/ Fishers' Least Significant Difference Test at p=0.05; applicable to comparisons between plot types (sprayed vs. landfarmed) and rates of application.

Conclusions. An experimental design for determining under field conditions the degradation, translocation, and phytotoxicity of herbicide contaminants in landfarmed waste has been presented. The major objective of the design was to provide adequate experimental units for comparing herbicide behavior after spraying with herbicide behavior after landfarming. Preliminary results of initial herbicide residues in soil and in runoff have been analyzed to determine the feasibility of the design. By analyzing six individual soil cores per replicate plot, variance of the mean herbicide concentration between replicate plots was lowered to levels that could be reasonably expected for a field study. Runoff collectors had enough capacity to collect storm runoff of 5.5 cm or less. Differences in concentrations of herbicides in runoff correlated positively with initial rate of application, but total herbicide runoff as a percentage of initial application seemed independent of increasing herbicide loads. Herbicide concentrations in runoff water were probably not high enough to cause significant toxicity in the algal photosynthetic inhibition bioassay. Soybean plant stand counts indicated that initial rates of application must be carefully controlled and made as low as feasible to avoid crop phytotoxicity, especially when a mixture of herbicide contaminants are present. Evidence of weed control confirmed that the herbicide residues in landfarmed soil were biologically active. Future reports will compare the persistence of herbicide residues in sprayed and landfarmed plots.

Acknowledgements

K. Dzantor, D. Kimme, L. Case, V. Chorney, S. Maddock, and J. Lang provided technical assistance. This research is a contribution from the Illinois Natural History Survey and Illinois Agricultural Experiment Station, College of Agriculture, University of Illinois at Urbana-Champaign, and was supported in part by the Illinois Hazardous Waste Research and Information Center, Project No. HWR 91-084, and the Illinois Fertilizer and Chemical Association.

Literature Cited

1. Habecker, M. A. *Environmental contamination at Wisconsin pesticide mixing/loading facilites: case study, investigation and remedial action evaluation.* Wisconsin Department of Agriculture, Trade, and Consumer Protection Agricultural Resource Division, Madison, WI, **1989**; 80 pp.
2. Long, T. *Proc. Illinois Agricultural Pesticides Conf. '89.* Coop. Ext. Serv. Univ. of Ill., Urbana-Champaign, IL, **1989;** pp. 139-149.
3. Enlow, P. D. *A Literature Review of Nonbiological Remediation Technologies Which May be Applicable to Fertilizer/Agrichemical Dealer Sites;* TVA Bulletin Y-213; Tennessee Valley Authority National Fertilizer & Environmental Research Center: Muscle Shoals, AL, **1990;** 20 pp.
4. Norwood, V. M.; Randolph, M. E. *A Literature Review of Biological Treatment and Bioremediation Technologies Which May Be Applicable at Fertilizer/Agrichemical Dealer Sites.;* TVA Bulletin Y-215; Tennessee Valley Authority National Fertilizer & Environmental Research Center: Muscle Shoals, AL, **1990;** 34 pp.
5. Loehr, R. M.; Overcash, M. R. *J. Environ. Engineering*, **1985**, *111*, 141-159.
6. Soil Science Society of America. *Utilization, treatment, and disposal of waste on land.* Soil Science Society of America, Inc., Madison, WI, **1986;** 318 pp.
7. Felsot, A. S. *Proc. 9th Ann. Hazardous Materials Management Conference/International*; Tower Conference Management Co.: Glen Ellyn, IL, **1991**; pp 506-515.
8. Bicki, T.; Felsot, A. S. In *Mechanisms of Movement of Pesticides into Groundwater;* Honeycutt , R. C., Schabacker, D., Ed.; ACS Symposium Series, American Chemical Society, Washington, D.C. **1992** (in press).
9. Crawford, R. L.; Mohn, W. W. *Enzyme Microb. Technol.*, **1985**, *7*, 617-620.
10. McGinnis, G. D.; Borazjani, H.; McFarland, L. K.; Pope, D. F.; Strobel, D. A. *Characterization and laboratory soil treatability studies for creosote and pentachlorophenol sludges and contaminated soil.* U.S. EPA report no. 600/S2-88/055. R. S. Kerr Environmental Res. Lab., Ada, OK, **1989.**
11. Valo, R.; Salkinoja-Salonen, M. Appl. Microbiol. Biotechnol., **1986,** *25,* 68-75.
12. Mueller, J. G.; Lantz, S. E.; Blattmann, B. O.; P. J. Chapman. *Environ. Sci. Technol.* ***1991,*** *25,* 1045-1055.
13. Wilson, G. B.; Sikora, L. J.; Parr, J. F. In *Land Treatment of Hazardous Wastes;* Parr, J. F.; Marsh, P. B.; Kla, J. M., Ed; Noyes Data Corp.: Park Ridge, NJ, **1983**; pp 263-273.
14. Kearney, P. C.; Muldoon, M. T.; Somich, C. J.; Ruth, J. M.; Voaden, D. J. *J. Agric. Food Chem.*, **1988,** *36*, 1301-1306.
15. Somich, C. J.; Kearney, P. C.; Muldoon, M. T.; Elsasser, S. *J. Agric. Food Chem.*, **1988,** *36*, 1322-1326.
16. Winterlin, W. L.; McChesney, M. M.; Schoen, S. R.; Seiber, J. N. *J. Environ. Sci. Health,* **1986,** *B21*, 507-528.
17. Winterlin, W.; Seiber, J. N.; Craigmill, A.; Baier, T.; Woodrow, J.; Walker, G. *Arch. Environ. Contam. Toxicol.*, **1989**, *18*, 734-747.
18. Dzantor, E. K.; Felsot, A. S. *Proc. International Wordshop on Research in Pesticide Treatment/Disposal/Waste Minimization*, U.S. EPA: Cincinnati, Ohio, **1992** (in press).

19. Felsot, A. S. In *1991 Illinois Fertilizer Conference Proceedings*; Hoeft, R. G., Ed.; Univ. of Ill. Cooperative Extension Service, Urbana-Champaign, Ill, **1991**; pp. 45-59.
20. Mullins, D. E.; Young, R. W.; Hetzel, G. H.; Berry, D. F. *Proc. 9th Ann. Hazardous Materials Management Conference/International*; Tower Conference Management Co.: Glen Ellyn, IL, **1991**; pp 503-505.
21. Felsot, A. S.; Liebl, R.; Bicki, T; *Feasibility of land application of soils contaminated with pesticide waste as a remediation practice*. Final Project Report HWRIC RR-021; Hazardous Waste Research and Information Center: Savoy, IL, **1988**; 55 pp.
22. Felsot, A.; Dzantor, E. K.; Case, L.; Liebl, R. *Assessment of problems associated with landfilling or land application of pesticide waste and feasibility of cleanup by microbiologial degradation*. Final Project Report HWRIC RR-053; IL Hazardous Waste Research & Information Center: Champaign, IL, **1990**; 68 pp.
23. Felsot, A. S.; Dzantor, E. K.; In *Pesticides in the Next Decade: the Challenges Ahead*; Weigmann, D. L., Ed; Proc. Third National Research Conference on Pesticides; Va. Polytechnic Inst.: Blacksburg, VA, **1991**; pp 532-551.
24. Felsot, A. S.; Mitchell, J. K.; Kenimer, A. L.; *J. Environ. Qual.*, **1990,** *19*, 539-545.
25. Ross, P. E.; Jarry, V.; Sloterdijk, H. In *ASTM STP 988*; Cairns, Jr.; Pratt, J. R., Eds.; American Society for Testing and Materials: Philadelphia, PA, **1988**; pp 68-73.
26. Felsot, A. S.; Dzantor, E. K. In *Enhanced Biodegradation of Pesticides in the Environment*; Racke, K. D.; Coats, J. R., Eds; Am. Chem. Soc. Symp. Ser. No. 426, Am. Chem. Soc.: Washington, D. C., **1990;** pp 249-268.
27. Taylor, A. W.; Freeman, H. P.; Edwards, W. M. *J. Agric. Food Chem.*, **1971**, *19*, 832-836.
28. Walker, A.; Brown, P. A. *Crop Protection*, **1983**, *2*, 17-25.
29. Hall, J. K.; Pawlus, M.; Higgins, E. R. *J. Environ. Quality*, **1972**, *1*, 172-176.
30. Roseboom, D.; Felsot, A.; Hill, T.; Rodsater, J. *Stream yields from agricultural chemicals and feedlot runoff from an Illinois watershed*. Final Project Report ILENR/RE-WR-90/11; IL Department of Energy & Natural Resources: Springfield, IL, **1990**; 133 pp.
31. Frank, R.; Braun, H. E.; Holdrinet, M. V. H.; Sirons, G. J.; Ripley, B. D. *J. Environ. Qual.*, *11*, **1982**, 497-505.
32. Mayasich, J. M.; Kalander, E. P.; Terlizzi, D. E., Jr.; *Aqualtic Toxicology*, **1987,** *10*, 187-197.

RECEIVED May 15, 1992

Author Index

Affiliation Index

Subject Index

E

I

J

K

L

M

R

S

Production: Paula M. Bérard, Margaret J. Brown, and Betsy Kulamer
Indexing: Deborah H. Steiner
Acquisition: A. Maureen Rouhi and Anne Wilson
Cover design: Sarah Sonyong Chung

Printed and bound by Maple Press, York, PA

Highlights from ACS Books

Good Laboratory Practices: An Agrochemical Perspective
Edited by Willa Y. Garner and Maureen S. Barge
ACS Symposium Series No. 369; 168 pp; clothbound, ISBN 0–8412–1480–8

Silent Spring Revisited
Edited by Gino J. Marco, Robert M. Hollingworth, and William Durham
214 pp; clothbound, ISBN 0–8412–0980–4; paperback, ISBN 0–8412–0981–2

Insecticides of Plant Origin
Edited by J. T. Arnason, B. J. R. Philogène, and Peter Morand
ACS Symposium Series No. 387; 214 pp; clothbound, ISBN 0–8412–1569–3

Chemistry and Crime: From Sherlock Holmes to Today's Courtroom
Edited by Samuel M. Gerber
135 pp; clothbound, ISBN 0–8412–0784–4; paperback, ISBN 0–8412–0785–2

Handbook of Chemical Property Estimation Methods
By Warren J. Lyman, William F. Reehl, and David H. Rosenblatt
960 pp; clothbound, ISBN 0–8412–1761–0

The Beilstein Online Database: Implementation, Content, and Retrieval
Edited by Stephen R. Heller
ACS Symposium Series No. 436; 168 pp; clothbound, ISBN 0–8412–1862–5

Materials for Nonlinear Optics: Chemical Perspectives
Edited by Seth R. Marder, John E. Sohn, and Galen D. Stucky
ACS Symposium Series No. 455; 750 pp; clothbound; ISBN 0–8412–1939–7

Polymer Characterization:
Physical Property, Spectroscopic, and Chromatographic Methods
Edited by Clara D. Craver and Theodore Provder
Advances in Chemistry No. 227; 512 pp; clothbound, ISBN 0–8412–1651–7

From Caveman to Chemist: Circumstances and Achievements
By Hugh W. Salzberg
300 pp; clothbound, ISBN 0–8412–1786–6; paperback, ISBN 0–8412–1787–4

The Green Flame: Surviving Government Secrecy
By Andrew Dequasie
300 pp; clothbound, ISBN 0–8412–1857–9

For further information and a free catalog of ACS books, contact:
American Chemical Society
Distribution Office, Department 225
1155 16th Street, NW, Washington, DC 20036
Telephone 800–227–5558